The Strategy of Culture
including
The Military Implications of the American Constitution
and
Roman Law and the British Empire

By

Harold Adams Innis

Cover Photograph: Luiz Fernando / Sonia Maria

ISBN: 978-1-78139-061-0

© 2012 Oxford City Press, Oxford.

Contents

PREFACE i
THE STRATEGY OF CULTURE 1
THE MILITARY IMPLICATIONS OF THE
    AMERICAN CONSTITUTION 23
    I 23
    II 26
    III 43
    IV 48
ROMAN LAW AND THE BRITISH EMPIRE 53

PREFACE

These two essays ["The Strategy of Culture" and "The Military Implications of the American Constitution"] assume a familiarity with a general thesis developed in *The Bias of Communication* (Toronto, 1951), in the Cust Foundation Lecture *Great Britain, the United States and Canada* (Nottingham, 1948), the Stamp Lecture, *The Press, a Neglected Factor in the Economic History of the Twentieth Century* (London, 1948) and a sesquicentennial lecture of the University of New Brunswick, *Roman Law and the British Empire* (1950).

The general argument has been powerfully developed in Aeschylus, *Prometheus Bound* as outlined in E. A. Havelock, *The Crucifixion of Intellectual Man* (Boston, 1950). Intellectual man of the nineteenth century was the first to estimate absolute nullity in time. The present—real, insistent, complex, and treated as an independent system, the foreshortening of practical prevision in the field of human action—has penetrated the most vulnerable areas of public policy. War has become a result and a cause of the limitations placed on the forethinker. Power and its assistant force, the natural enemies of intelligence, have become more serious since "the mental processes activated in the pursuit and consolidating of power are essentially short range" (p. 99). But it will not do to join the great chorus of those who create a crisis by saying there is a crisis.

H. A. I.

THE STRATEGY OF CULTURE

WITH SPECIAL REFERENCE TO CANADIAN
LITERATURE—A FOOTNOTE TO THE MASSEY REPORT

"Pay them well; where there is a Maecenas there will be a Horace and a Virgil also." *Martial*

"Complaints are made that we have no literature; this is the fault of the Minister of the Interior." *Napoleon*

The title of this article may be regarded as an illustration of the remark of Julien Benda concerning "the *intellectual organization of political hatreds*" [1] and as a further effort to exploit Canadian nationalism. "Political passions rendered universal, coherent, homogeneous, permanent, preponderant—everyone can recognize there to a great extent the work of the cheap daily political newspaper." [2] Whistler[3] and others have contended that art is not to be induced by artificial tactics. They have pointed to Switzerland as a country without art and it has interesting parallels with Canada, a country of more than one language, a federation, and dependent on the tourist trade. A distinguished Canadian painter has remarked: "I am not sure that future opinion of the contemporary art of our day will not consider the advertising poster, the window and counter card as most representative." [4]

Printers' ink threatens to submerge even the literary arts in Canada and it may seem futile to raise the question of cultural possibilities.

---

[1] Julien Benda, The Great Betrayal (London, 1928), p. 21.
[2] Ibid., p. 7.
[3] J. M. Whistler, The Gentle Art of Making Enemies (New York, 1904).
[4] William Colgate, C. W. Jeffreys (Toronto, n.d.), p. 28.

THE STRATEGY OF CULTURE

The power of nationalism, parochialism, bigotry, and industrialism may seem too great. Cheap supplies of paper produce pulp and paper schools of writing, and literature is provided in series, sold by subscription, and used as an article of furniture. Almost alone Stephen Leacock, by virtue of his mastery of language, escaped into artistic freedom and was recognized universally and even he, as Peter McArthur pointed out, never attacked a publisher.

But we can at least point to the conditions which seem fatal to cultural interests. We can appraise the cultural level of the United States and appreciate the importance of New York as a centre for the publication of books and periodicals, the effects of the higher costs of commercial printing in Chicago, and the dangers to literature and the drama of reliance on the authoritative finality of New York newspaper critics. We should be able to escape the influence of a western American news agency which advised that if you want it to sell "put a New York date line on it."

We can point to the dangers of exploitation through nationalism, our own and that of others. To be destructive under these circumstances is to be constructive. Not to be British or American but Canadian is not necessarily to be parochial. We must rely on our own efforts and we must remember that cultural strength comes from Europe. [1] We can point to our limitations in literature and to the consequent distortions incidental to the impact of mechanization, notably in photography. The story has been compelled to recognize the demands of the illustration and has become dominated by it. [2] The impact of the machine has been evident in the dependence of Edgar Wallace and Phillips Oppenheim

---

[1] "Until the English visitor to America comprehends that he is in the midst of a civilization totally different from anything he has known on our side of the Atlantic, he is exposed to countless shocks." Sir John Pollock, Bt., Time's Chariot (London, 1950), pp. 184-5. Sir John regards the great difference as having developed since 1880 as a result of the Civil War and foreign immigration. In England, with a background of feudalism, it seems possible to keep political differences and personal relationships in separate departments.

[2] Whistler's complaint that painting was subordinate to literature must be offset by the account of Newman Flower of Cassell & Co. He resorted to a cliché department or "bank" of illustrations built up since 1870, selected a promising illustration, and asked a young writer to write around it. Just As It Happened (London, 1951), p. 27.

and dictators of the quick action novel on the dictaphone. [1] An emphasis on speed and action essential to books produced for individual reading weakens the position of poetry and the drama particularly in new countries swamped by print.

Burckhardt[2] in his studies of Western civilization held that religion and the state were stable powers striving to maintain themselves and that civilized culture did not coincide with these two powers, that in its true nature it was actually opposed to them. "Artists, poets and philosophers have just two functions, i.e. to bring the inner significance of the period and the world to ideal vision and to transmit this as an imperishable record to posterity." In the words of Sir Douglas Copland, summarizing the philosophy of P. H. Roxby, "A cultural heritage is a more enduring foundation for national prestige than political power or commercial gain." [3] "It is the cultural approach of one nation to another, which in the long run is the best guarantee for real understanding and friendship and for good commercial and political relations. In the past, it has been, on the whole, sadly neglected, and especially as between western Europe and China." (Roxby.) [4] It has been scarcely less neglected as between Canada and the United States. In the long list of volumes of "The Relations of Canada and the United States" series, little interest is shown in cultural relations and the omission is ominous.

Inter-relations between American and Canadian publishing in the nineteenth century had significant implications for Canadian literature in the present century. In the nineteenth century the tyranny of the novel in England had been built up in part because of inadequate protection to English playwrights from translations of French plays,

---

[1] Ibid., p. 40. On the other hand Edgar Wallace protested that dictaphone stuff was "good Wallace publicity. I write my best stuff with a pen." Reginald Pound, Their Moods and Mine (London, 1939), p. 233. "Dictation always is rubbish" (George Moore). Ibid., p. 112. As a result of the influence of the newspaper on reading, novels have been written to be read rapidly and consequently emphasize length and description. "I do not want literature in a newspaper" (E. L. Godkin).

[2] See Jacob Burckhardt, Force and Freedom: Reflections on History (New York, 1943).

[3] D. B. Copland, "Culture versus Power in International Relations" in Liberty and Learning: Essays in Honour of Sir James Hight (Christchurch, 1950), p. 155.

[4] Ibid., p. 154.

production of which had been systematically encouraged in France, [1] and by a monopoly of circulating libraries protected by the high price of the three-volume novel which made it, therefore, cheaper to rent than to buy books. [2] Restrictive effects of high prices on exports of books from Great Britain, absence of circulating libraries in the United States, lack of protection to foreign, especially English books before the enactment of copyright legislation in America in 1891, and section 5 of the American Copyright Act, May 31, 1790, which was "an invitation to reprint the work of English authors," were factors responsible for large-scale reprinting of English works in the United States and for the publication of English works first in the United States. [3]

In 1874 legislation in the United States reduced postage on newspapers issued weekly or oftener to two cents a pound without regard to the distance carried. Under an act of March 3, 1879 (par. 14), second-class mail matter "must be regularly issued at stated intervals as frequently as four times a year, and bear a date of issue, and be numbered consecutively." Again, on July 1, 1885, postal charges on paper-covered books were reduced from two cents per pound to one cent and cloth-bound books were carried at eight cents per pound. The legislation reflected the demands of a vigorous cheap book publishing period, concentrating on English or foreign books for which a market had been created by established publishers.

In the ultimate development of the publication of English books previous to the Copyright Act in 1891, Canadians, emigrants to the United States and undisciplined by the demands of its distributing machinery, played an important role. George Munro, a mathematics teacher in the Free Church College, Halifax, who had emigrated to New York and acquired experience in the handling of dime novels in the firm of Beadle and Adams and in the publishing of the *Fireside Companion*, a family newspaper started in 1867, launched the "Sea-

---

[1] In France the Théâtre Français was subsidized by the government, and the Society of Dramatic Authors founded by Beaumarchais and reorganized by Scribe in the nineteenth century fostered an interest in plays rather than novels. See Brander Matthews, Gateways to Literature and Other Essays (New York, 1912), p. 41 and also H. A. Innis, Political Economy in the Modern State (Toronto, 1946), pp. 35-55.
[2] See introduction by Graham Pollard to I. R. Brussel, Anglo-American First Editions, 1826-1900 (New York, 1935), p. 10.
[3] Ibid., p. 11. See also H.A. Innis, The Bias of Communication (Toronto, 1951), pp. 171-2.

side Library," a quarto, two or three columns to the page with cheap paper, on May 28, 1877. It was estimated that 645 pages in a regular edition could be printed in 152 pages quarto. As a result of saturation of the market for quartos in the latter part of 1883, Munro started a pocket-size edition in spite of the higher costs of manufacturing. In 1887 he cut wholesale prices from twenty and twenty-five cents to ten cents and from ten cents to five cents, and in 1889 sought protection by publishing a monthly "Library of American Authors," cheap cloth-bound twelvemos, "sold by the ton." In 1890 Munro sold the "Seaside Library" to J.W. Lovell, [1] on a three-year option to repurchase arrangement, for $50,000 plus $4,500 monthly. It was estimated that, by 1890, 30,000,000 volumes of the "Seaside Library" had been sold, chiefly through the American News Company.

J. W. Lovell was the son of John Lovell, who in 1872 had a printing shop on the American side near Montreal at which he printed British copyright works, free of copyright, and imported them into Canada under 12½ per cent duty to be sold at a lower price than editions imported from Great Britain. [2] The son moved to New York in 1875 and engaged in the sale of cheap unauthorized editions. After a failure in 1881 he followed the German plan of producing cheap handy books with neat covers and, in 1882, started publication of handy twelvemos in "Lovell's Library," paper-covered books selling at twenty cents, and "Lovell's Standard Library," cloth-bound at one dollar. In 1885 he concentrated on "Lovell's Library" and sold the remainder of his business to Belford and Clarke. This became a most popular series selling about seven million volumes annually. As a result of the reduction of prices by George Munro in 1887 competition became more intense and in 1888 Lovell bought the "Munro Library" [3] from Norman W. Munro, the brother of George Munro. The "Munro Library" in pocket-size books had been started in 1884 when the owner had returned to the business after failing with the "Riverside Library," sold between 1877 and 1879. With control over the "Seaside Library," acquired in 1890, and over the plates and stock of other cheap book publishers by purchase or rental to the extent of over half the titles of

---

[1] The funds became the basis of a substantial gift to Dalhousie University.
[2] R. H. Shove, Cheap Book Production in the United States, 1870 to 1891 (Urbana, 1937), p. 75. This book is a mine of information.
[3] This included 855 sets of plates and 1,500,000 copies of books for which $250,000 was paid.

cloth-bound books and over three-fourths of the titles of paper-covered books, and supported by the Trow Printing Company, Lovell organized the United States Book Company with a reported capital of $3,500,000.

Alexander Belford and James Clarke, members of a firm of Belford brothers in Toronto, moved to Chicago and organized Rose, Belford and Company; it was reorganized in 1879 after a failure as Belford Clarke and Company. They became publishers of "railroad literature" and built up an elaborate retail system developing a policy of selling to the book trade at artificially high prices, first to jobbers, and then to the regular trade, and later at extremely low prices through dry-goods and department stores. Showy bindings contrasted with the woodpulp, clay, and straw paper inside the books. In 1885 they acquired "Lovell's Standard Library" and became the largest producers of cheap cloth-bound twelvemos. As a result of the intensive price cutting after 1887 they failed in 1889.

In the absence of copyright on foreign books, publishers were compelled to rely on their only means of protection, namely, cheapness based on mass production. With efficient systems of distribution through the American News Company and the post office, equipment was steadily improved; cylinder presses were first installed in 1882 and in 1886 three cheap library publishers had their own typesetting, printing, and binding plants. The cheapest variety of paper was used and slight attention was given to proof reading and corrections. Paper manufacturers were compelled to sell their fine book papers chiefly to the large printing houses and the periodical publishers. Stereotype establishments or "sawmills" began to sell plates to publishers who then issued their own editions. Typographical unions [1] complained, and, following the sharp reduction in prices, recognized the importance of copyright. With lower postal rates on paper-covered editions, and prices from one-sixth to one-tenth those of cloth-bound volumes, it was estimated that almost two-thirds of a total of 1,022 books published in 1887 were issued in the cheap libraries. Demands for new titles led to the publication of poorer classes of fiction. [2] The technological

---

[1] The unions were at first opposed to the Copyright Act but became active in its support; see G. A. Tracy, History of the Typographical Union (Indianapolis, 1913), p. 450.
[2] Brussel, Anglo-American First Editions, p. 19.

changes which lowered the prices of paper [1] and of printing widened the gap between the supply of written material and the demand of readers and intensified the need for non-copyright foreign books. Yet the supply of foreign material was limited, the market for lower grade fiction was saturated, it was no longer possible to increase sales by changing formats from quarto to twelvemo, deterioration of paper was not sufficiently rapid, and finally newspapers expanded to absorb supplies of newsprint. Publishers were now compelled to emphasize American writers, to whom copyright was paid. The basis was laid for the supremacy of the periodical, with significant consequences for American and Canadian literature. National advertising steadily advanced to impose its demands on the reading material of the periodical. The discrepancy between prices of books in England and in the United States gradually lessened. The three-volume novel disappeared in England as prices were levelled with those in the United States after the Copyright Act of 1891. To secure copyright it was necessary to print books in the United States. [2]

In the last decade of the nineteenth century the advantages of cheap newsprint, of cheap composition following the invention of the linotype, and of the fast press as the basis of large circulations were being fully exploited by newspapers. Every conceivable device to increase circulation was pressed into service, notably in the newspaper war between Pulitzer and Hearst in the late nineties in New York City, including sensational headlines, the comics, and the Spanish American War. Crusades were started in every direction to enhance goodwill for newspapers.

The sudden improvement in technology in the production of newspapers was accompanied by an increase in magazine readers. The weekly was replaced by the monthly which became a leading factor in modern publishing. The Copyright Act of 1891, in itself a recognition

---

[1] In 1871, newsprint straw paper was twelve cents per pound, fine book paper sixteen to seventeen cents; in 1875 newsprint was nine cents, machine-finish book paper ten to eleven cents; in 1889 newsprint was three and one-quarter cents and calendared book paper six and one-half to seven and one-half cents. Shove, Cheap Book Production, p. 4.

[2] Cheap unauthorized editions disappeared and the works of authors such as Kipling, which had sold widely in pirated editions, were sold at higher prices and in smaller numbers.

of the problem of creating a supply of American writers, [1] was followed by the training of an army of fiction writers who by 1900 met the demands of magazines. Muck-raking magazines [2] were supported by experienced newspaper men such as Lincoln Steffens (who wrote a series on "The Shame of the Cities"). They followed the tactics, particularly of the Hearst newspapers, in the struggle for circulation. [3]McClure, for instance, applied the sensational methods of the cheap newspaper to the cheap and new magazine. He sponsored a reform wave which was effectively exploited by Theodore Roosevelt. He built up circulation by paying enormous sums to famous writers and trying to corner a market in them. As a former peddler of coffee pots, he knew the demands of people on farms and in small towns. [4] Munsey, [5] in the all-fiction magazine which followed the Sunday magazine section of the newspaper with smooth paper and clearer half-tones, made fiction the basis of circulation and earning power by 1896. [6]

The position of women as purchasers of goods led to concentration on women's magazines and on advertising. In Philadelphia, Curtis developed the great discovery, that reading matter trailed through a periodical compelled readers to turn the pages and to look at the advertising which made up most of the page, into an extensive magazine business. [7] Through the national magazine, [8] advertisers such as the manufacturers of pianos, high cost two-wheeled bicycles, and other commodities were able to reach a large market at less cost than through the daily newspaper and to concentrate on more attractive layouts appealing to people in higher income brackets. The national

---

[1] The suit brought against the New York World by Harriet Monroe for printing her ode presented at the opening of the Chicago World's Fair and the award of $5,000 damages strengthened the position of authors. Harriet Monroe, A Poet's Life: Seventy Years in a Changing World (New York, 1938), pp. 139-43.
[2] C. C. Regier, The Era of the Muckrakers (Chapel Hill, N.C., 1932).
[3] H. L. Mencken, Prejudices, First Series (New York, 1929), p. 175.
[4] S. S. McClure, My Autobiography (New York, 1914).
[5] F. A. Munsey, The Founding of the Munsey Publishing-House (New York, 1907); also George Britt, Forty Years—Forty Millions: The Career of Frank A. Munsey (New York, 1935).
[6] Algernon Tassin, The Magazine in America (New York, 1916), pp. 342-3.
[7] Arthur Train, My Day in Court (New York, 1939), p. 419.
[8] Frank Presbrey, The History and Development of Advertising (New York, 1929), p. 339.

magazine made a systematic attack on older advertising media. Religious papers dependent on patent medicine advertising felt the effects of a crusade of the *Ladies' Home Journal* which in 1892 [1] refused to handle medical advertising and exposed widely advertised preparations by printing chemical analyses. With the growth of large-scale printing, the printer assumed the direction of advertising and displaced the single advertiser and agency. Specialization of printing and increased pressure of overhead costs necessitated effective control of publications. Lorimer, an able writer of advertisements, became editor of the *Saturday Evening Post* and gave advertisements the personality of articles. [2] A four-colour printing press costing $800,000 and a new building in 1910 led the Curtis publications to add a third magazine to cover agriculture. [3]

The average circulations of magazines increased from 500,000 to 1,400,000 in the period from 1905 to 1915 and following the boom beginning in 1922 reached 3,000,000 by 1937. [4] The *Reader's Digest* was started in 1922, *Time* in 1923, and the *New Yorker* in 1925. Extension of education and increased use of text-books conditioned youth to acceptance of the printed word and to magazine consumption. The demand for writers exceeded the supply. After the First World War, women's magazines, which had begun as pattern makers in the *Delineator* and other Butterick papers, gained conspicuously in circulation. Women's magazines reached the largest circulations, paid most highly for articles, and were the chief market for writers. Competition between magazines for writers with an established reputation brought

---

[1] Ibid., pp. 531-2. See The Americanization of Edward Bok: The Autobiography of a Dutch Boy Fifty Years After (New York, 1937). Also Edward W. Bok, A Man from Maine (New York, 1923). The campaign against patent medicines provoked the announcement by Eugene Field of the engagement of the grand-daughter of Lydia W. Pinkham to Edward W. Bok, the editor of the Ladies' Home Journal.

[2] Bok, A Man from Maine, p. 171. "The secrets of success as an editor were easily learned; the highest was that of getting advertisements. Ten pages of advertising made an editor a success; five marked him as a failure." The Education of Henry Adams: An Autobiography (Boston, 1918), p. 308. "The art of advertising has outgrown the art of creative writing .... Three-fourths of the income of the magazines comes from their advertisers ... just take the advertising and rewrite it." W. E. Woodward, Bunk (New York, 1923), p. 51.

[3] Bok, A Man from Maine, p. 183.

[4] Train, My Day in Court, p. 421.

sky-rocket prices.¹ The sale of film rights to popular novels brought even more than that of serial rights. An average best-seller in "the slicks" with serial rights, movie, book, and other rights brought returns varying between $70,000 and $125,000. Writers concentrated on magazines rather than books.²

Writing for the great popular magazines built up on advertising implied assiduous attention to their requirements on the part of writers and editors. Dullness was absolutely abhorrent. Serial instalments involved consideration of appropriate terminal points at which intense interest might be sustained for the next number. Magazines with the largest circulation were able to carry longer fiction by writers with an established reputation but tended to reduce instalments and stories from 12,000 to 5,000 or 4,500 words.³ Since dependence on advertising meant that the magazine "expands and contracts with the activity of the factory chimney"⁴ writers were particularly affected by fluctuations of the business cycle. The reputations of authors were built up through advertising by editors of magazines who were thus enabled to sell advertising material, and stories⁵ became commercialistic. George Ade could write "I guess I can now sell anything I write, even if it's good."⁶

The influence of the newspaper and advertising on the magazine was developed to a sophisticated level in the twenties when magazines such as the *New Yorker* playfully exposed the foibles of its advertisers and advertisers exploited the foibles of the magazine. More recently the campaign of the *New Yorker* against loud speaker advertising in public buildings has not been unrelated to competition for advertis-

---

¹ Fairfax Downey, Richard Harding Davis, His Day (New York, 1933), p. 219.
² Ibid., pp. 430-1, 433.
³ Train, My Day in Court, pp. 423-5. In England Gilbert Frankau held that the serial market was disappearing because readers of monthly magazines would not wait and newspapers preferred the short story "in these days of so much front-page excitement." Pound, Their Moods and Mine, p. 241.
⁴ Train, My Day in Court, p. 420. The limited circulation of Canadian magazines makes for a seasonal expansion. Advertising is sufficient only during the period of the two or three months before Christmas to warrant a full-fledged interest in features, especially short features. Longer features appear after the holiday season.
⁵ Train, My Day in Court, p. 440.
⁶ F. W. Wile, News is Where You Find It (Indianapolis, 1939), p. 36.

ing—all of course in the spirit of good clean fun. The rigid limitations in style of advertising copy enabled the *New Yorker* to succeed by emphasizing the independence of the editor from the business office, and by developing a new style of writing which in turn led to a revolution in the style of advertising copy. In the *Smart Set* and the *American Mercury* H. L. Mencken, a Baltimore newspaperman, was successful in building up circulation in a direct attack on the limitations of a society dependent on advertising. In reviewing books for newspapers he had become familiar with trends in literature and he attracted to the *Smart Set* new authors unable to secure publication with old firms and willing to acquire prestige in lieu of high rates of pay. As a columnist Mencken had also gained an intimate knowledge of libel laws. Of German descent, he had suffered from the frenzied propaganda of the First World War. The *American Mercury* was started in 1924 as a fifty-cent magazine and practically doubled its average monthly circulation from 38,694 to 77,921 by 1926. [1] Debunking became a new word and a profitable activity. In developing the *American Mercury* as a quality magazine designed to make the common man respectable, [2] Mencken pursued his attacks on the puritanical and on the English book to the point of recognizing in a powerful fashion the new language of the newspaper and the magazine in his *American Language*.

The women's magazines began to feel the restraining influence of puritanism and its effects on advertising. Bok became concerned with the importance of sex education. Theodore Dreiser, editor of *Delineator*, came into conflict with censorship regulations in his novels and triumphantly conquered in *An American Tragedy*. Mencken, in the tradition of Mark Twain and Ambrose Bierce, secured the support of the Authors' League for Dreiser's position. [3] The Calvinistic obsession of hypocritical people with the subject of sex [4] became the centre of attack by Dreiser as chief artist and Mencken as high priest, determined to defeat "the iron madonna who strangles in her fond embrace the American novelist" (H. H. Boyesen). With a shrewd appreciation of the advertising value of censorship regulations Mencken seized upon the occasion of the banning of a copy of the *American Mercury* to

---

[1] W. Manchester, Disturber of the Peace: The Life of H. L. Mencken (New York, 1951), p. 15.
[2] Ibid., p. 155.
[3] Ibid., pp. 93-4.
[4] Ibid., p. 101.

attack the Boston Watch and Ward Society as the stronghold of Catholic and Protestant puritanism. [1] His active interest in the Scopes trial, following a law enacted in Tennessee on March 21, 1925, against the teaching of evolution was a part of the general strategy against religious bigotry.

Decline of the practice of reading aloud led to a decline in the importance of censorship. The individual was taken over by the printing industry and his interest developed in material not suited to general conversation. George Moore in England and H. L. Mencken in the United States exploited the change in their attacks on censorship. Censorship could no longer be relied upon to secure publicity. Significantly the advertiser had contributed to a change of atmosphere and women no longer feared to smoke cigarettes in public.

Even before the Copyright Act, the effects of advertising, as reflected in the newspaper and the magazine, on the writer had important implications for the book. "Most people now do not read books, but read magazines and newspapers" (H. G. Baird). [2] Limited distributing facilities for books evident in the high costs of book agents and subscription publishing [3] in the nineties, and the development of special publishers of text-books in the early part of the century were gradually being offset by department stores. Small retail stores for books could not compete with rents paid by diamonds, furs, and bonds. Mail order business in books expanded in the early 1900's but the results were perhaps evident in the remark of a publisher's reader, "this novel is bad enough to succeed." [4] W. D. Howells wrote in 1902: "Most of the best literature now sees the light in the magazines, and most of the second

---

[1] Ibid., p. 207. See an account of the failure of attempts by Covici, Friede to secure suppression of Radclyffe Hall's Well of Loneliness by the Boston Watch and Ward Society. Donald Friede, The Mechanical Angel (New York, 1948), p. 94.
[2] J. C. Derby, Fifty Years among Authors, Books and Publishers (New York, 1884), p. 559.
[3] Subscription selling was accompanied by a development of techniques of salesmanship and depended for its success to an important extent on snob appeal, particularly the prestige attached to owning a large book among the relatively illiterate. Estes and Lauriat of Boston, prominent subscription book agents, who came under the control of Walter Jackson and Harry E. Hooper after 1900, were active in developing schemes for the sale of the Encyclopaedia Britannica in connection with the London Times.
[4] W. H. Page, A Publisher's Confession (New York, 1905), p. 27.

best appears first in book form." The increasing importance of apartment buildings and lack of space for shelves supported the rapid development of the lending library in the twenties. Book clubs increased rapidly [1] after 1926 as a means of securing the economies of mass production. Nevertheless, the inadequacy of book distributing machinery and dependence on British and Continental devices [2] showed the limitations of the book in contrast with the newspaper and the magazine. Publishing firms such as Doubleday, Page and Company entered on policies of direct vigorous advertising, which built up, for instance, the success of O. Henry, [3] but their most significant results were in less obvious directions.

The experience of the prominent publishing firm of Scribner's illustrates directly the impact of advertising on the newspaper and the magazine and in turn on the book. Roger Burlingame, [4] trained in a newspaper office, and M. E. Perkins, a reporter on the *New York Times*, exercised a powerful influence on publications of the firm. Perkins was concerned to arouse a consciousness of the value and importance of the native note in opposition to the imitation of English and European models and "the cynical disparagement of American materialism." [5] To him great books were those which appealed to both the literati and the masses. The book-buying public was made up of fairly successful people but to Perkins the reading of Thomas Wolfe's books "to pieces" in the libraries reflected the truer sense of life of people in the lower economic level. [6] While he condemned the mad pursuit of best-sellers which developed during the boom period of the twenties and the newspaper policy of playing up the work of authors of best-sellers and criticized the Book of the Month Club for concentrating the attention of the public on one book a month, [7] he was concerned primarily with the newspaper public. Writers from the newspaper field included Hemingway, Edmund Wilson, Stanley Pennell, Stephen Crane, and Dreiser. It was his opinion that the teaching

---

[1] E. H. Dodd, The First Hundred Years: A History of the House of Dodd, Mead, 1839-1939 (New York, 1939), p. 36.
[2] O. M. Sayer, Revolt in the Arts (New York, 1930).
[3] Train, My Day in Court, p. 439.
[4] Roger Burlingame, Of Making Many Books (New York, 1946), p. 221.
[5] J. H. Wheelock, Editor to Author: The Letters of Maxwell E. Perkins (New York, 1950), p. 8.
[6] Ibid., p. 184.
[7] Ibid., p. 128.

of literature and writing in the colleges compelled students to see things through a film of past literature and not with their own eyes. Two years with a newspaper were better than two years in college. [1] He favoured what Irving Babbitt called "art without selection." The demands of commercialism were evident more directly in the avoidance of controversy. "The sales department always want a novel. They want to turn everything into a novel." [2] The public and the trade preferred books of 100,000 words and works of 25,000 to 30,000 words were padded to give the appearance of books of a larger size.

An orderly revolt against commercialism was significantly delayed and frustrated in literature possibly more than in any other art. Henry James had escaped to England and in the period after the First World War Ezra Pound and T. S. Eliot followed. "The historians of Wolfe's era ... all record this strange phase of our cultural adolescence; the same sad and distraught search for foreign roots." [3] "You could always come back" (Hemingway). But in the words of Pound: "We want a better grade of work than present systems of publishing are willing to pay for." [4] "The problem is *how*, how in hell to exist without overproduction." [5] "The book-trade, accursed of god, man and nature, makes no provision for *any* publication that is not one of a series ...." [6] "The American law as it stands or stood is all for the publisher and the printer and all against the author, and more and more against him just in such proportion as he is before or against his time." [7] Books by living authors were, he claimed, kept out of the United States and "the tariff, which is iniquitous and stupid in principle, is made an excuse." [8] Even in Great Britain from about 1912 to 1932 booksellers did "their

---

[1] Ibid., p. 267. "What the eighteenth century thought simply vulgar, and the nineteenth gathered data from, has now become literary material; even the annals of the poor are to be short and simple no longer." H. W. Boynton, Journalism and Literature and Other Essays (Boston, 1904), p. 164.
[2] Wheelock, Editor to Author, p. 84.
[3] Maxwell Geismar, Writers in Crisis: The American Novel between Two Wars (Boston, 1942), pp. 214 and passim.
[4] The Letters of Ezra Pound, 1907-1941, ed. D. D. Paige (New York, 1950), p. 175.
[5] Ibid., p. viii.
[6] Ibid., p. 319.
[7] Ibid., p. 52.
[8] Ibid., p. 53. See J. L. May, John Lane and the Nineties (London, 1936), p. 159.

utmost to keep anything worth reading out of print and out of ordinary distribution." "Four old bigots" of Fleet Street practically controlled the distribution of printed matter in England.[1] Criticism was related to publishers' advertising.[2]

The distorting effects of industrialism and advertising on culture in the United States have been evident on every hand. Architecture as a sort of tyrant of the arts had the advantage of the utilitarian demands of commerce. Painting and sculpture as allied to it had the support of collectors, private and public, and the encouragement of awards and prizes.[3] Poetry was the subject of paragraphers' jokes, a space filler for magazines[4] and "must appeal to the barber's wife of the Middle West."[5] "Poetry had no one to speak for it."[6] In the drama the lack of interest of actors in modern art[7] and the support of tradition involved effective reliance on Shakespeare and a terrific handicap to playwrights.[8] The commercial theatre manager and the newspaper critic have been reluctant to recognize the vitality of a demand for the imaginative artistic work of the little theatre[9] particularly in competition with the cinema. In the words of George Jean Nathan the talking picture may be "the drama of a machine age designed for the consumption of robots" and the theatre may have gained enormously by the withdrawal of "shallow and imbecile audiences," but the change has been costly and painful.[10]

---

[1] Letters of Pound, pp. 239-40.
[2] Ibid., p. 337.
[3] Harriet Monroe, A Poet's Life, p. 241.
[4] Ibid., p. 247. A study of the demands of space on Bliss Carman's poetry might prove rewarding.
[5] Ibid., p. 288.
[6] Ibid., p. 242.
[7] Nathan refers to "the mean capacity of the overwhelming number of them, whatever their nationality ... the downright ignorance, often made so conspicuously manifest." The Intimate Note-books of George Jean Nathan (New York, 1932), p. 144.
[8] See a letter from Mrs. Fiske in Harriet Monroe, A Poet's Life, pp. 176-7.
[9] Ibid., p. 419.
[10] See St. John Ervine, The Alleged Art of the Cinema (n.p., March 15, 1934). "Actors and actresses were certainly regarded with far greater interest than they are nowadays. The outstanding ones inspired something deeper than interest. It was with excitement, with wonder and with reverence, with something akin even to hysteria, that they were gazed upon. Some of the younger of you listeners would, no doubt, if they could, interrupt me at this point by

THE STRATEGY OF CULTURE

The overwhelming pressure of mechanization evident in the newspaper and the magazine has led to the creation of vast monopolies of communication. Their entrenched positions involve a continuous, systematic, ruthless destruction of elements of permanence essential to cultural activity. The emphasis on change is the only permanent characteristic. Thomas Hardy complained that narrative and verse were losing organic form and symmetry, the force of reserve, and the emphasis on understatement, and becoming structureless and conglomerate.[1]

The guarantee of freedom of the press under the Bill of Rights in the United States and its encouragement by postal regulations has meant an unrestricted operation of commercial forces and an impact of technology on communication tempered only by commercialism itself.[2] Vast monopolies of communication have shown their power in securing a removal of tariffs on imports of pulp and paper from Canada though their full influence has been checked by provincial governments especially through control over pulpwood cut on Crown lands. The finished product in the form of advertisements and reading material is imported into Canada with a lack of restraint from the federal government which reflects American influence in an adherence to the principle of freedom of the press and its encouragement of monopoly.

---

asking, 'But surely you don't mean, do you, that our parents and grandparents were affected by them as we are by cinema stars?' I would assure you that those idols were even more ardently worshipped than are yours. Yours after all, are but images of idols, mere shadows of glory. Those others were their own selves, creatures of flesh and blood, there, before our eyes. They were performing in our presence. And of our presence they were aware. Even we, in all our humility, acted as stimulants to them. The magnetism diffused by them across the footlights was in some degree our own doing. You, on the other hand, have nothing to do with the performances of which you witness the result. Those performances—or rather those innumerable rehearsals—took place in some far-away gaunt studio in Hollywood or elsewhere, months ago. Those moving shadows will be making identically the same movements at the next performance or rather at the next record; and in the inflexions of those voices enlarged and preserved for you there by machinery not one cadence will be altered. Thus the theatre has certain advantages over the cinema, and in virtue of them will continue to survive." Sir Max Beerbohm in The Listener, Oct. 11, 1945, p. 397.

[1] May, John Lane and the Nineties, p. 177.

[2] See Upton Sinclair, Money Writes! A Study of American Literature (Long Beach, Calif., 1927).

Sporadic attempts have been made to check this influence in Canada as in the case of the banning of the Hearst papers in the First World War and in the imposition by the Bennett administration of a tariff based on advertising content in American periodicals. Protests are made by institutions against specific articles in American periodicals but without significant results other than that of advertising the periodical. To offset possible handicaps Canadian editions of *Time*, *Reader's Digest* and the like are published. Canadians are persistently bombarded with subscription blanks soliciting subscriptions to American magazines, and their conversation shifts with regularity following the appearance of new jokes in American periodicals. Canadian publications supported by the advertising of products of American branch plants and forced to compete with American publications imitate them in format, style and content. Canadian writers must adapt themselves to American standards. [1] Our poets and painters are reduced to the status of sandwich men. The ludicrous character of the problem may be shown by stating that the only effective means of sponsoring Canadian literature involves a rigid prohibition against all American periodicals with any written material and free admission to all periodicals with advertising only. In this way trade might be fostered and Canadian writers left free to work out their own solutions to the problems of Canadian literature. Indeed they would have the advantage of having access to the highly skilled examples of advertiser's copy.

Publishers' lists in Canada are revealing in showing the position of American branches or American agencies in the publication of books. Advertising rates for a wide range of commodities, determined by newspapers and magazines particularly in relation to circulation, are such as to make it extremely difficult for publishers to compete for advertising space, particularly as book advertising is largely deprived of the powerful force of repetition. [2] Moreover, the demands of a wide range of industries for advertising compete directly and effectively for raw materials, paper, capital, and labour entering into the production of books, and restrict the possibility of advertising them. American devices such as book clubs and the mass production of pocket books to

---

[1] One Canadian writer has complained of writing an article of 60,000 words for an American woman's magazine, cutting it to about 40,000 words to make two instalments, and expanding it to 80,000 for the English market. Canadian writers should become efficient concertina players.

[2] Wheelock, Editor to Author, p. 138.

be sold on news-stands and in cigar stores and drug stores have immediate repercussions in Canada. The extreme importance of book titles—perhaps the most vital element in American literature—evident in the changing of titles of English books in the United States and of American books in Great Britain and in the interest of the movie industry in the publishing field, [1] is felt in Canada also. In the field of the newspaper, dependence on the Associated Press and other agencies, on the *New York Times*, [2] and other media needs no elaboration. In radio and in television accessibility to American stations means a constant bombardment of Canadians.

The impact of commercialism from the United States has been enormously accentuated by war. Prior to the First World War the development of advertising[3] stimulated the establishment of schools of commerce and the production of text-books on the psychology of advertising. European countries were influenced by the effectiveness of American propaganda. Young Germans were placed with American newspaper chains and advertising and publishing agencies to learn the

---

[1] See J. T. Farrell, The Fate of Writing in America (n.p., n.d.), also W. T. Miller, The Book Industry (New York, 1949). "Before the war British publishers were often told by friends in the Canadian book trade that their public preferred the bigger, handsomer American book. They wanted value for money, and had been accustomed to measure value by size and weight. The story has often been told of the Canadian agent who handed one of his travellers an advance copy of a new book from a British publisher and asked, 'How many can you sell of that?' The traveller, without opening the book, handed it back and said, 'None.' The agent, somewhat nettled, said, 'None? But you haven't even looked at it.' The traveller replied, 'I don't need to. It doesn't weigh enough.'" Michael Joseph, The Adventure of Publishing (London, 1949), p. 131.

[2] It "set out to be dull and ponderous and it has achieved its purpose with a fidelity and thoroughness justly commanding the admiration of all lovers of bulk and solidity." G. M. Fuller, "The Paralysis of the Press," American Mercury, Feb., 1926, p. 160.

[3] Will Irwin, Propaganda and the News (New York, 1936). For an account of the influence of an advertising agent of a Canadian department store on advertising and journalistic ideas in England, see Autobiography of a Journalist edited with an introduction by Michael Joseph (London, n.d.), pp. 45, 50. The author, advised by the agent to begin journalism by writing advertisements for shopkeepers, used samples of full-page advertisements of the Canadian store (p. 66). Advertising methods were then introduced effectively in political campaigns.

art of making and slanting news. American treatises on advertising and publicity were imported and translated. American graduate students were attracted to Germany by scholarships and experiments in municipal government. In turn, German exchange professorships were established, especially with South American universities. The Hamburg-American Lines became an effective propagandist organization.

But German experience [1] proved much too short in contrast with that of American [2] and English propagandists, [3] though their effectiveness is difficult to appraise since the estimates have been provided chiefly by those responsible for the propaganda.

American propaganda [4] after the First World War became more intense in the domestic field. Its effectiveness was evident in the emergence of organizations representing industry, labour, agriculture, and other groups. The Anti-Saloon League pressed its activities to success in prohibition legislation. In the depression the American government [5] learned much of the art of propaganda from business and exploited new technological devices such as the radio. With the entry of the United States into the Second World War instruments of propaganda [6] were enormously extended.

The effects of these developments on Canadian culture have been disastrous. Indeed they threaten Canadian national life. The cultural life of English-speaking Canadians subjected to constant hammering from American commercialism is increasingly separated from the cultural life of French-speaking Canadians. American influence on the

---

[1] G. S. Viereck, Spreading Germs of Hate (New York, 1930).
[2] James R. Mock and Cedric Larson, Words That Won the War: The Story of the Committee on Public Information, 1917-1919 (Princeton, N.J., 1939).
[3] See Neville Lytton, The Press and the General Staff (London, 1921); Sir Campbell Stuart, Secrets of Crewe House: The Story of a Famous Campaign (London, 1920); Walter Millis, Road to War: America 1914-1917 (Boston, 1935); James Squires, British Propaganda at Home and in the United States from 1914 to 1917 (Cambridge, Mass., 1935); H. D. Lasswell, Propaganda Technique in the World War (London, 1927).
[4] See O. W. Riegel, Mobilizing for Chaos: The Story of the New Propaganda (New Haven, Conn., 1939).
[5] See George Michael, Handout (New York, 1935); L. C. Rosten, The Washington Correspondents (New York, 1937).
[6] See Propaganda by Short Wave ed. H. L. Childs and J. R. Whitton (Princeton, N.J., 1943); C. J. Rolo, Radio Goes to War: The "Fourth Front" (New York, 1940).

latter is checked by the barrier of the French language but is much less hampered by visual media. In the period from 1915 to 1920 the theatre in French Canada was replaced by the movie or French influence by American. With the development of the radio, protection of language enabled French Canadians to take an active part in the preparation of script and in the presentation of plays. During the Second World War the revue and the French-Canadian novel received fresh stimulus. The effects of American technological change on Canadian cultural life have been finally evident in the numerous suggestions of American periodicals that Canada should join the United States. It should be said that this would result in greater consideration of Canadian sentiment by American periodicals than is at present the case when it probably counts for less than that of a religious sect.

The dangers to national existence warrant an energetic programme to offset them. In the new technological developments Canadians can escape American influence in communication media other than those affected by appeals to the "freedom of the press." The Canadian Press has emphasized Canadian news but American influence is powerful. [1] In the radio, on the other hand, the Canadian government in the Canadian Broadcasting Corporation has undertaken an active role in offsetting the influence of American broadcasters. It may be hoped that its role will be even more active in television. The Film Board has been set up and designed to weaken the pressure of American films. The appointment and the report of the Royal Commission on National Development in the Arts and Sciences imply a determination to strengthen our position. The reluctance of American branch plants to support research in Canadian educational institutions has been met by taxation and federal grants to universities. Universities have taken a zealous interest in Canadian Literature but a far greater interest is needed in the whole field of the fine arts. Organizations such as the Canadian Authors' Association have attempted to sponsor Canadian literature by the use of medals and other devices. The resentment of English and French Canadians over the treatment of a French-

---

[1] "I am sceptical about the value of 90 per cent of press reports. Most of them tend to say enough to be misleading and not enough to be in any sense informative." Interview with a veteran Vancouver journalist. See M. L. Ernst, The First Freedom (New York, 1946) and Herbert Brucker, Freedom of Information (New York, 1949).

Canadian play on Broadway points to powerful latent support for Canadian cultural activity.

We are indeed fighting for our lives. The pernicious influence of American advertising reflected especially in the periodical press and the powerful persistent impact of commercialism have been evident in all the ramifications of Canadian life. The jackals of communication systems are constantly on the alert to destroy every vestige of sentiment toward Great Britain holding it of no advantage if it threatens the omnipotence of American commercialism. This is to strike at the heart of cultural life in Canada. The pride taken in improving our status in the British Commonwealth of Nations has made it difficult for us to realize that our status on the North American continent is on the verge of disappearing. Continentalism assisted in the achievement of autonomy and has consequently become more dangerous. We can only survive by taking persistent action at strategic points against American imperialism in all its attractive guises. By attempting constructive efforts to explore the cultural possibilities of various media[1] of communication and to develop them along lines free from commercialism, Canadians might make a contribution to the cultural life of the United States by releasing it from dependence on the sale of tobacco and other commodities which would in some way compensate for the damage it did before the enactment of the American Copyright Act.

---

[1] The problem to an important extent centres around the confusion as to the distinct possibilities of each medium. Literary agents deliberately exploit the demands of technological innovations, adapting the same artistic piece of work to the book, the magazine, and the film. See Curtis Brown, Contacts (London, 1935). Shaw refused to allow a play to be filmed stating that no one would go to see it after seeing it on the screen and that the author suffered because the play became dull with the dialogue left out (ibid., p. 51). The studios wanted "a big kick" at the end of every sequence of the film (ibid., p. 33). Mechanization demands uniformity. The newspapers are concerned with news and contemporary topics, and books, plays, films, and novels centre around newspaper owners. The book has been subordinated to the demands of advertising for the movies, business firms in centennial volumes, radio broadcasts, and articles from magazines. Bible scenes are exploited for plays and movies. Shakespeare's plays for actors are primarily studied in print as texts. Newspaper serials and radio scripts differ from novels and emphasize topics of the widest general interest. Any fresh idea is immediately pounced on and mauled to death. Irvin Cobb remarked concerning the dull conversation of Hollywood that the phrase coiners preserved silence until they had sold the wheeze themselves.

THE MILITARY IMPLICATIONS OF THE AMERICAN CONSTITUTION

I

This paper [1] is an attempt to understand the policies of the United States. In Canada we are under particular obligations to attempt such an understanding in our own interests as well as in the interests of the rest of the world. The difficulties involved in any country's understanding itself, particularly a country with a complex unstable history, are overwhelming and the most penetrating studies of the United States have been made by de Tocqueville, a Frenchman, and by Lord Bryce, an Englishman. A Canadian is too close to make an effective study but he has the most to gain from it. He is handicapped by tradition especially in English-speaking Canada, evident in the pervasive influence of those who left the United States after the American Revolution, namely the United Empire Loyalists, and by language in French-speaking Canada. The writer of this paper can scarcely pretend to the necessary objectivity, nor, I suspect, can most of his readers. Nevertheless we must do our best.

Whatever our view about the American Revolution we must agree that it was achieved by a resort to arms against Great Britain. To the British it may have been a war of little consequence; we remember the remarks of an Englishman who when told that in the War of 1812 the British forces had burned Washington said he thought he had died in bed. To Americans the achievement was a result of desperate struggle. Revolutions leave unalterable scars and nations which have been burned over by them have exhibited the most chauvinistic brand of

---

[1] Read at a meeting of the Salmagundi Club on December 6, 1951.

nationalism and crowd-patriotism. [1] These nations have developed highly depersonalized social relationships, political structures, and ideals and their counsels are determined most of all by spasms of crowd propaganda. "Public policy sits on the doorstep of every man's personal conscience. The citizen in us eats up the man." [2] The founders of the American Constitution appear to have recognized the danger by framing an instrument which put limits on the number of things concerning which a majority could encroach on the position of the individual. [3] But the extent of such protection has varied and declined with improvements in the technology of communication and the increasing powers of the executive, as Senator McCarthy has conspicuously shown.

Washington and his successors in the nineteenth century renounced an interest in Europe but steadily expanded their influence in the Americas following the increase in demand for new land on which to raise cotton. The demand implied steady expansion westward, in the south, and, in order to maintain a balance, in the north. In the south expansion was at the expense of the French empire, notably in Jefferson's administration when Louisiana was bought from Napoleon, and in the north at the expense of the British empire when Lewis and Clark were sent on a journey of exploration to the northwest and when John Jacob Astor established Astoria on the Columbia River. Later expansion in the south was safeguarded in the Monroe Doctrine, enunciated in 1823, which warned European powers to keep their hands off South America and was directed to the absorption of Texas, California, and other states at the expense of the Spanish empire and of Mexico. The remnants of a crumbling Spanish empire were finally taken over after the explosion of the *Maine* in Cuba ("Remember the *Maine*") and when Puerto Rico and the Philippines became American possessions. Expansion in the south to some extent intensified and to some extent

---

[1] E. D. Martin, The Behavior of Crowds: A Psychological Study (New York, 1920), p. 223.
[2] Ibid., p. 248.
[3] Ibid., p. 249. "The most certain test by which we judge whether a country is really free, is the amount of security enjoyed by minorities." "By liberty I mean the assurance that every man shall be protected in doing what he believes his duty, against the influence of authority and majorities, custom and opinion .... It is bad to be oppressed by a minority, but it is worse to be oppressed by a majority." (Lord Acton.) See Sir John Pollock, Bt., Time's Chariot (London, 1950), pp. 166-7.

eased the pressure on the British empire in the north. The line was eventually tightened to the present Canadian border and Alaska, "Seward's icebox," was purchased from Russia in 1867. These developments remind us of Disraeli's comment when Poland had been partitioned by European powers at a meeting at breakfast. "What will they have for lunch?"

II

The outbreak of the American Revolution marked a return to ideological warfare such as had largely disappeared in England after the Civil War.[1] Democratic nationalism and the mass army became the new basis of warfare. [2] George Washington, an officer in the British army in the Seven Years' War against the French, had gained experience which gave him the leadership of the Revolutionary Army. The immediate significance of the Revolution was evident in the position of this soldier from Virginia. A mass army could not be built up under a New England general. [3] As a result of success in arms he secured not only independence for the colonies but also a stable federal government. He presided over the Convention and was asked to take the chief position in the new government. An interest in western lands was not unrelated to his sympathy with the Federalists in their proposal for a strong central government with "powers competent to all general purposes," words included in a letter from him to Hamilton in 1783. [4] His sympathies found reflection in the views of delegates concerned about the dangers implicit in the radical character of state constitutions written by revolutionary legislatures. "Our chief danger rises from the democratic parts of our constitutions" (Edmund Randolph of Virginia to the Convention). [5] Conservatism and an emphasis on the theory of divided powers led to provisions strengthening the executive power,

---

[1] J. F. C. Fuller, Armament and History (London, 1946), p. 101.
[2] Ibid., p. 109.
[3] Herbert Agar, The United States: The Presidents, the Parties & the Constitution (London, 1950), p. 28. "For it is a fact, that more than one third of their general officers have been inn-keepers, and have been chiefly indebted to that circumstance for such rank. Because by that public, but inferior station, their principles and persons became more generally known." Smyth; cited by Kittredge, The Old Farmer and His Almanack (Cambridge, Mass., 1920), p. 264.
[4] Agar, The United States, p. 37.
[5] Ibid., p. 45.

such as those making the President Commander-in-Chief of the Army and Navy and giving him control over patronage. The Secretaries of State and War were made responsible to the President alone and, with the exception of the Treasury Department, the precedent was followed in the establishment of new Cabinet posts. The President became a focus of executive power. The influence and character of Washington finally left their impression on the United States as he secured Virginia's acceptance of the Constitution in 1787 and gave leadership to the other states which followed.

In the work of establishing a nation, the influence and prestige of the first President left an indelible impression on the operation of government. However, Washington's efforts to secure the advice of the Senate as a sort of privy council were met with distrust. The decision of the Senate to receive reports of Cabinet ministers in writing and to exclude them from its meetings drove the Cabinet into the position of being the President's council. As a further guarantee against presidential interference, in Congress a system of committees was emphasized in which members were protected by secrecy from any group including the press.

John Adams, the second President (1797-1801), whose election implied a recognition of the role of New England in the Revolution and its aftermath, inherited the task of maintaining the prestige of the office, but he found it difficult to maintain the delicate balance between New England and the South, in the face of the power of Alexander Hamilton as a representative of industrial and commercial interests in the middle states. At Hamilton's insistence, Washington had agreed to call out the militia of four states to put down the Whisky Rebellion in 1794. In 1798 Hamilton advised his friends in the government to prepare for war with France, and Congress planned for a large emergency army and an increase in the regular army. Under his influence Washington agreed to head the army and by virtue of his prestige could insist on choosing his generals. Strife between Adams and Hamilton was followed by defeat of the former for a second term and by a weakening of the Federalist position.

In opposition to the centralizing tendencies of the Constitution, Jefferson (1801-9) led a group whose views were reflected in the Declaration of Independence and the Articles of Confederation. He emphasized the position of the land, the small farmer, and the labourer against banking and the commercial interests. On his trip up the Hudson with Madison in 1791 he laid the foundations for the "longest-lived, the most incongruous, and the most effective political alliance in

American history: the alliance of southern agrarians and northern city bosses." [1] In contrast with the Federalists who insisted that survival depended on the sword, Jefferson stated: "I hope no American will ever lose sight of the essential policy of interdicting in the seas and territories of both Americas, the ferocious and sanguinary contests of Europe." "Our first and fundamental maxim should be, never to entangle ourselves in the broils of Europe." [2] As a representative of the South, and in spite of his statement that "our peculiar security is in the possession of a written Constitution," he accepted the annexation of Louisiana and acquired the port of New Orleans without asking the question of constitutional propriety. To an alliance between the city bosses of New York and the South, he added the West.

After Jefferson's two terms, Madison, also a native of Virginia, became President (1809-17) and acquired additional territory. On April 14, 1812, Congress formally divided West Florida at the Pearl River, annexing the western half to the new state of Louisiana, and, a month later, the eastern half to the Mississippi Territory. In 1813 the American army forced the Spanish garrison at Mobile to surrender and took possession. Henry Clay and the Committee on Foreign Affairs persuaded Congress to declare war on Great Britain on June 18, 1812. "The conquest of Canada is in your power." "This war, the measures which preceded it, and the mode of carrying it on, were all undeniably Southern and Western policy, and not the policy of the commercial states" (Josiah Quincy). [3] On December 5, 1814, Madison recommended liberal spending on the Army and the Navy and the establishment of military academies.

Following the two terms of Madison, Monroe, again a native of Virginia, and an officer in the Revolutionary Army, became President (1817-25). The decline of the Federalist party meant that there was no official opposition, and also no party discipline. The President was thus left without any device to secure cohesion in Congress. In the House of Representatives, for example, an Army bill, opposed by the President and the Secretary of War, was "carried notwithstanding many defects in the details of the bill by an overwhelming majority." [4]

---

[1] Ibid., p. 88.
[2] Washington, of course, in his Farewell Address had said, "It is our true policy to steer clear of permanent alliances with any portion of the foreign world, so far, I mean, as we are now at liberty to do it."
[3] Cited Agar, The United States, p. 174.
[4] Ibid., p. 200.

In 1822 Monroe recognized the independence of the Latin American republics which had been part of the Spanish empire, and, on the insistence of John Quincy Adams, included in his statement of the Monroe Doctrine on December 2, 1823, a protest against the encroachment of Russians in the northwest.

The success of the War of 1812 and the re-election of Monroe in 1820 finally destroyed the Federalist party as a political factor. Decline in prestige and power of the congressional caucus opened the way for a free fight in 1824; New England influence was once more reflected in the election of John Quincy Adams, who like his father, John Adams, served only for one term (1825-9).

His successor, Andrew Jackson (1829-37), a native of South Carolina, had suffered at the hands of the British in the Revolutionary War. In the War of 1812 he had led western militiamen against the Indians of Georgia and Alabama and destroyed British troops under General Sir Edward Pakenham in New Orleans. In 1817 he pursued marauding Indians into Spanish territory, marched to Pensacola, and removed the Spanish governor. After his invasion of Florida he became military governor. As a national figure and a popular hero he introduced a system of military organization to national politics. Beginning in 1825 he built up a national political machine. A small, divided, virulent, and undisciplined [1] press which had contributed to the disappearance of the Federalist party and a monopolistic Washington press were replaced by an organized party press designed to provide discipline and propaganda. The *National Intelligencer*, [2] the organ of Jefferson, Madison, Monroe, and J. Q. Adams, had been the oracle of war sentiment before and after 1812 and had a wide circulation for daily, semi-weekly, and weekly editions. [3] In opposition, Jackson and his followers established media to maintain a close contact with voters. After his election the

---

[1] James Cheetham, an exile from England after the Manchester riots in 1798, attempted in the American Citizen, a daily sponsored by Clinton, to break the power of Aaron Burr in New York. William Duane, editor of the powerful Jeffersonian paper, the Aurora, because of a bitter grudge against Madison and Gallatin who refused to give him a job contributed to the defeat of the Navigation Act of Gallatin and hastened the outbreak of war.

[2] This had been the Independent Gazetteer of Philadelphia under Joseph Gales, a son of the editor and proprietor of the Sheffield Register, who had left England following a charge of sedition in 1795. It was purchased by S. H. Smith in 1800 and moved to Washington.

[3] A. K. McClure, Recollections of Half a Century (Salem, 1902), pp. 37-9.

United States Telegraph and the Washington *Globe* became administrative mouthpieces for partisan purposes. [1] Rewards were offered to strengthen the morale of the troops; "no plunder no pay." Political organizers in state politics such as Van Buren at Albany were brought to the national stage. In 1832 at the time of the nomination of Jackson for a second term, a system of nominating conventions was introduced in which a two-thirds rule was invoked to protect the position of the South. The news value of the system became evident in the emergence of the presidential candidate as the chief consideration of politics. Under Jackson and his successor, Van Buren (1837-41), a representative of New York State, campaign techniques were elaborated. Veto messages, written up by journalistic members of the Kitchen Cabinet for popular consumption, had a wide distribution. The difficulties of the system became evident when attempts were made to meet the demands of regional groups. The Tariff of Abominations, and the opposition to Vice-President Calhoun of South Carolina in the nullification controversy, made the latter a defender of state rights and led to the enactment of the Force Act by which the President was given authority to call out the Army and Navy to enforce laws of Congress. The dragon's teeth of secession were sown.

To meet the type of organization built up in support of Jackson and Van Buren, an attempt was made to establish a Whig Party, based chiefly on anti-Masonic feeling, [2] following the contest of 1836. In New York State, Seward and Weed, to weaken the position of Van Buren and to exploit the news value of a war hero, secured the nomination of W. H. Harrison, who had been engaged in a battle with the Indians at Tippecanoe Creek in 1811, and was promoted to command the Army of the Northwest in the War of 1812. A vigorous campaign with an emphasis on such slogans as "log cabin and hard cider" led to his election in 1841 but his death shortly afterwards meant the elevation of the vice-president, Tyler, a native of Virginia. Texas, which had seceded from Mexico in 1836, was annexed to the United States near the end of his administration (1841-5), and formally admitted on

---

[1] J. E. Pollard, The Presidents and the Press (New York, 1937), p. 147.
[2] The anti-Masonic party put Seward in the New York State Senate in 1830, made Joseph Ritner Governor in Pennsylvania in 1835, and supported an alliance of J. Q. Adams, William Wirt, Francis Granger, and Thurlow Weed. It carried Vermont for Wirt and Ellmaker, candidates for President. C. T. Congdon, Reminiscences of a Journalist (Boston, 1880), p. 29.

July 4, 1845. The Texas issue defeated Clay's hopes of the presidency in 1844 and weakened the Whig party.

J. K. Polk (1845-9), a native of North Carolina, the first dark horse ever nominated for the presidency, aggressively pressed for settlement of the Oregon boundary dispute under the slogan "Fifty-Four Forty or Fight" and secured recognition of a boundary in 1846. This aggressiveness was designed to increase the number of states in the north, to parallel the increase in the south with the addition of Texas and the acquisition of New Mexico and California. Americans in California took a hint from Polk and declared an independent state. Polk ordered General Zachary Taylor to occupy the left bank of the Rio Grande; at length the exasperating Mexicans committed an overt act, which was followed by a brief successful war. In 1847, in "the spot" resolutions, Lincoln took an active part in attacking Polk, and to a resolution of Congress thanking General Taylor, secured the addition of a rider that the war had been started by Polk "unnecessarily and unconstitutionally." [1] Polk [2] was accused by the Whigs of forcing a war to extend the institution of slavery. Opposition to the aggressiveness of the south in the interests of new territory became more vocal through the activities of Lincoln and organs such as the *Chicago Tribune*.

Again to capture the electorate, Thurlow Weed, a skilful journalist and politician, played an active role in securing the selection of General Taylor, a native of Virginia, and the hero of Buena Vista (February 1847). He was selected at the Iowa convention within a month of his victory and later triumphantly elected. Vice-President Fillmore, a native of New York, became President on his death in 1850 and like most vice-presidents not in harmony with the policy of the administration, reversed it. He was sympathetic to the South, and made the first effort of a president to purge his party by opposing the nomination of Whig congressmen who had voted against the Clay compromise. [3] In 1852 the Whigs nominated Winfield Scott, the general who had led the troops to Mexico City, but he was defeated by Pierce (1853-7). Newspapers exploited such remarks of Scott as "I never read the *New York Herald*" and "the hasty plate of soup."

---

[1] See R. S. Harper, Lincoln and the Press (New York, 1951), p. 9.
[2] T. W. Barnes, Memoir of Thurlow Weed (Boston, 1884), p. 172.
[3] H. L. Stoddard, Horace Greeley, Printer, Editor, Crusader (New York, 1946), p. 149.

The long struggle between the North and South was drawing to a close as the North was no longer able to offset southern influence by such tactics as nominating generals for President. These tactics had been to an extent self-defeating since military power was reinforced by recognition of heroes in elections to the presidency. The Whig party [1] was replaced by the Republican party supported by the free soil movement. The plantation system led to the acquisition of Indian and Mexican lands. The spoils of Mexico were poisoning the political system—each addition of territory accentuated the rivalry between North and South. The gold rush in California precipitated a more intense struggle for control over the first transcontinental railway. Jefferson Davis, Secretary of War under Pierce, a native of New Hampshire and a minor national hero at Buena Vista, insisted on a Pacific railway along the Mexican border linking California to the Gulf states and opening the trade of Asia to the plantation society. In the north, on the other hand, Stephen Douglas of Illinois demanded a route through Nebraska.

Mastery of the South was evident in the nomination and election of weak northern presidents—Pierce and Buchanan (1857-61), the latter with the advantage of having refused to wear court dress in England, [2] and the distinction of being the only president from Pennsylvania. Compromises between North and South included the reciprocity treaty with the British colonies in 1854 designed to extend the influence of the North as a balance to expansion in the South. Finally the Supreme Court reflected the influence of the South when it appeared as an agent for southern expansion in the Dred Scott decision. The nomination of Lincoln from the Middle West by the Republican party and his election brought southern expansionism to an end. Robert E. Lee, a contemporary of Jefferson Davis at West Point, became in 1865 General-in-Chief of the Confederate armies. Withdrawal of able generals to the southern armies compelled the North to build up the effectiveness of a widely separated staff, with activities co-ordinated through the telegraph; the attempt was eventually successful under Grant. Inef-

---

[1] The Whigs failed to capture the popular vote. Daniel Webster was alleged to have said that they should "come down into the forum and take the people by the hand," words which were printed innumerable times in the largest type in Democratic newspapers. Governor J. A. Clifford, on the other hand, imprudently called the Democrats "poor in character and meager in numbers." Congdon, Reminiscences of a Journalist, p. 61.

[2] Pollard, The Presidents and the Press, p. 293.

ficient military leadership in the North meant a longer period of war, greater loss of life, and greater bitterness toward the South. After the savagery which characterized Sherman's march through Georgia to the sea, reflected in his remark "War is hell," the prospects of reconciliation were slight. A revival in the Civil War of the savagery of ideological warfare established precedents for the twentieth century.

At the end of the Civil War a national army had emerged to serve a national state. The President and executive were supreme above the states. Washington became the significant capital and state governments became less important. The South was invited to join a vastly different union than that she had left, but in turn the war had created a solid and a different South from the one which had left the union. Ideological warfare had been carried to great lengths. The North imposed a peace more bitter than war. The Republican party, as a result of the costs of civil war and victory, became a sacred cause to New England, the farmers of the Middle West, veterans concerned with pensions, and negroes. Andrew Johnson (1865-9) was finally disregarded as President. In spite of the Constitution, the President was deprived of control of the Army and governments in the South which had been elected in 1865 were replaced in 1867 by military rule with the whole area divided into five military districts each under a major general. Grant, trained as a general, became the head of an executive which had been built up by a skilful politician but which had deteriorated under Johnson who followed the precedent of vice-presidents in reversing policy. Like Jefferson Davis, Grant carried the dominating qualities of a soldier into the administration of civil affairs (1869-77). He was thwarted in his ambition to annex San Domingo in the south by Sumner, chairman of the Foreign Relations Committee of the Senate, who long served as a focus of northern bitterness, following the savage physical attack on him by Brooks of South Carolina on the floor of the Senate, [1] and who insisted on the acquisition [2] of Canada to the north.

With the aggressive support of Union veterans of the Grand Army of the Republic, Hayes, a brigadier general under Sheridan, was elected to the presidency by a narrow margin in 1876 (1877-81). In his fight with the Senate, the telegraph became an effective instrument in the mobilization of public opinion. He acquired control of the appointive power and "the long domination of the executive by the Congress

---

[1] See Congdon, Reminiscences of a Journalist, p. 253.
[2] The Education of Henry Adams: An Autobiography (Boston, 1918), p. 275.

was at an end" (H. J. Eckenrode). Grant had been unable to restore the South to white rule because of the Army and the bitterness following the war but under Hayes, as a result of the cohesiveness of white southerners in the Democratic party, the retreat of the North from the South was begun. It was finally ended in 1894 and the negro was left a third-class citizen, legally free, but deprived of his vote. On the other hand Hayes began the unfortunate precedent of using his power over federal troops to break strikes in West Virginia, Pennsylvania, and Maryland.

Hayes was followed by James A. Garfield, a brigadier general at Shiloh, who to become President defeated General Winfield Scott Hancock, a Union commander at Gettysburg, "a good man weighing 280 pounds" (W. O. Bartlett, in the *Sun*). Garfield, supported by Whitelaw Reid of the New York *Tribune*, had defeated Conkling and the New York *Herald* in the attempt in 1880 to nominate Grant for a third term. [1] The appointive powers conceded to Hayes led to a concern with the introduction of civil service reform but since domination of the Senate necessitated a rigid control over patronage, a strict merit system was impossible. Factors responsible for the murder of Lincoln, vicious personal bitterness, the war, disappearance of an interest in great causes, and the growth of the spoils system culminated in the assassination of Garfield, [2] the defeat of Blaine and the election of Cleveland, and the return of the Democratic party. (Arthur, Vice-President under Garfield, became President in 1881, but contrary to the usual practice did not change his policy.)

On its return to power in 1885 the Democratic party and its President, though relatively free from the hatreds exploited by the Republican party, was inexperienced and undisciplined. A forceful leader, Cleveland (1885-89, 1893-7) strengthened further the position of the executive in opposition to the Senate. He was defeated by his tariff message of December, 1887, and by Benjamin Harrison (1889-93), [3] a grandson of President William Henry Harrison elected in

---

[1] Pollard, The Presidents and the Press, pp. 480-6.
[2] Agar, The United States, p. 533.
[3] J. S. Clarkson, assistant Postmaster-General, a former teacher and journalist, is said to have distributed 38,000 post offices and to have secured the election of Harrison in opposition to Blaine. Herbert Quick, One Man's Life (Indianapolis, 1925), p. 220. "In numerous instances the post-offices were made headquarters for local party committees and organizations and the centers of partizan scheming."

1840, a great grandson of a signer of the Declaration of Independence, and the last of the aristocrats in American politics, and a brevet major at the end of the Civil War. The unpopularity of the McKinley Tariff and the depression contributed to the re-election of Cleveland as president in 1892. Inexperience and lack of discipline in the party, and continuation of the depression were to defeat him. Neglect of monetary reform and an emphasis on the tariff, incidental to the revival of southern influence, led to Bland's warning to Cleveland, in the "parting of the ways speech" in 1893, and a breach between eastern and western Democrats. The weakness of Cleveland in the party was not unrelated to various tactics designed to strengthen his position as President. Although a Democrat he followed the precedent of Hayes in sending federal troops to stop the Pullman strike in Chicago and destroyed the last vestiges of state sovereignty which had maintained the safety of commerce depended on the power of the state. [1] Richard Olney, [2] his Secretary of State, held "any permanent political union between a European and an American state unnatural and inexpedient"—a statement of interest to Canadians. He sent instructions of an inflammatory nature to the American minister in London regarding the dispute between Great Britain and Venezuela, and Cleveland sent a message to Congress which revived feelings of antagonism against Britain. The Navy was rehabilitated and Mahan's writings on naval power developed as an important influence.

The vigorous note to Great Britain was designed to attract Irish votes since the Democratic party in the North had been built up around

---

Party literature favorable to the post-masters' party, that never passed regularly through the mails, was distributed through the post-offices as an item of party service, and matter of a political character, passing through the mails in the usual course and addressed to patrons belonging to the opposite party, was withheld; disgusting and irritating placards were prominently displayed in many post-offices, and the attention of the Democratic enquirers for mail matter was tauntingly directed to them by the post-masters." (Cleveland.) Cited by Agar, ibid., p. 550.
[1] McClure, Recollections of Half a Century, p. 131.
[2] He threatened the World with application of a statute of January 30, 1799, in complaint of its influence on the conduct of Great Britain in relation to the Venezuelan controversy. J. L. Heaton, The Story of a Page (New York, 1913), pp. 112, 122.

the Irish American element in New York State. [1] The words "Rum, Romanism and Rebellion" used by a supporter of Blaine had contributed to the latter's defeat in 1884. [2] In turn the outcome of the election of 1888 had been influenced by a letter which Sackville-West, British Minister in the United States, was tricked into writing to the effect that the interest of Great Britain would be best served by the return of Cleveland. [3] In that election the charge of subservience of the Democratic party to the Southern Confederacy had been heard for the last time. In 1896 the free silver campaign of the West drove the gold standard Democrats in the East out of politics and weaker elements of the party came to the surface. [4]

As a nominee of the Democratic party reflecting the demands of the West for monetary reform, Bryan was defeated by W. J. McKinley (1897-1901) who had served as a private, and was a brevet major at the end of the Civil War. The war mania, developed over the Venezuela dispute, persisted and led to demands for war with Spain. This Congress declared in April, 1898. "McKinley had in part given in to public pressure, for fear of disrupting his party and losing the autumn elections." [5] "From the Rio Grande to the Arctic Ocean there should be but one flag and one country!" was the cry of Henry Cabot Lodge. Regarding the Philippines, McKinley decided that "there was nothing left for us to do but take them all, and to educate and uplift and civilize and christianize them," a process involving a long period of hostilities with the Filipinos. [6] The Hawaiian Islands were annexed, partly be-

---

[1] W. J. Abbott, Watching the World Go By (Boston, 1933), p. 74. J. Y. McKane, a Coney Island boss, failing to secure benefits from Cleveland, became very active in opposition to him. James L. Ford, Forty-Odd Years in the Literary Shop (New York, 1921), pp. 345-6.
[2] As Thomas Nast had done effective work as a cartoonist in the election of Grant, Bernard Gillam particularly with "The Tattooed Man" in Puck was effective in his support of Cleveland. Ford, Forty-Odd Years, p. 299. Conkling's refusal to support Blaine in the words "I am not in the criminal practice" gave weight to the attack.
[3] Abbott, Watching the World Go By, p. 103. Cleveland asked for his recall. This probably served as a counter move to a release of a story in England in 1887 of the possible purchase of the Maritimes by the United States which was cabled as a scoop for American papers.
[4] J. D. Whelpley, American Public Opinion (London, 1914), p. 18.
[5] Agar, The United States, p. 624.
[6] Ibid., p. 625.

cause they would be needed to defend the Philippines. In the peace treaty Puerto Rico was ceded by Spain.

During the war in Cuba, Theodore Roosevelt, God's gift to newspapermen, who had raised the Rough Riders, and, with the assistance of Richard Harding Davis as war correspondent, secured important space on the front pages of newspapers, became a centre of attention. [1] He was elected Governor of New York State, became Vice-President in McKinley's second term, and President (1901-9) on the latter's assassination. This was attributed to an incendiary press, particularly the writings of Bierce and the Hearst papers, which supported the Democratic party. [2] Such was the background for a belief in power for the central government; "I achieved results only by appealing over the heads of the Senate and House leaders to the people, who were the masters of both of us." [3] Cleveland gave out messages on Sunday evenings [4] to get more space in the Monday papers and Roosevelt exploited the practice following the development of Sunday papers by making important statements on Sunday and compelling the dull Monday papers to feature them. [5] He prepared speeches well ahead of time in order that they could be distributed to all newspapers before public delivery and the expenses of telegraphing them be avoided. [6] The interest of newspapers in his activities was a result of his sense of news, and of his concern with trust busting, which implied defeat of the International Paper Company as a trust, and lower prices of newsprint. "I took the canal zone and let Congress debate." The Panama had "a most just and proper revolution." [7] In spite of Congress he sent the United States fleet to the Pacific to impress the Japanese. Under pressure from Roosevelt the Canadian claim in the Alaska boundary dispute had been sacrificed. [8] Regarding the appointment of judges to the Supreme

---

[1] Commenting on Roosevelt's Rough Riders, Mr. Dooley wrote: "'Tis 'Th' Biography iv a Hero be Wan who Knows.' ... If I was him I'd call th' book 'Alone in Cubia.'" Elmer Ellis, Mr. Dooley's America (New York, 1941), p. 145.
[2] Abbott, Watching the World Go By, p. 139.
[3] Agar, The United States, p. 639.
[4] Pollard, The Presidents and the Press, p. 517.
[5] Abbott, The United States, p. 244.
[6] Oscar King Davis, Released for Publication: Some Inside Political History of Theodore Roosevelt and His Times, 1898-1918 (Boston, 1925), p. 102.
[7] Agar, The United States, p. 650.
[8] Ibid., p. 626.

Court, Roosevelt wrote: "he [a judge of the Court] is not in my judgment fitted for the position unless he is a party man, a constructive statesman ...." [1] His position was summed up in his statement: "... I did greatly broaden the use of executive power." [2]

In 1909 W. H. Taft, the nominee of President Roosevelt and the Republican party, became President (1909-13). He had been Governor of the Philippines from 1900 to 1904, Secretary of War after 1904 when he successfully reorganized work on the Panama Canal and was described as "an amiable island, completely surrounded by men who know exactly what they want." He attempted to secure the passage of a reciprocity treaty in 1911 but the attitude of President Roosevelt in the Alaska boundary dispute had done much to stimulate hostility leading to its defeat in Canada. The increasing power of the executive, following Hayes and Cleveland, was accompanied by the emergence of the Speaker as an important channel between the executive and Congress. T. B. Reed became the Speaker in 1889, when the Republicans captured both houses and the presidency; a continuous representative from Maine, he was responsible for a marked increase in the importance of the position. The weakness of the Democratic party, and the position of the Speaker, first in the case of Reed and then in the case of Cannon, in the Republican party, precipitated a revolt in the latter party in 1910. After that date, the Speaker was excluded from membership in the Rules Committee of the House and lost his power to appoint its Standing Committees. As a result the President had no one person with whom he could deal, and bitterness between factions of the Republican party led to the emergence of ex-President Roosevelt with a Progressive party and the election of President Wilson in 1912.

The election of President Wilson (1913-21) was not only a result of the difficulties of the Republican party but also of the steady improvement in the discipline and solidarity of the Democratic party. Champ Clark's blunder in coining a phrase which was used with such telling effect in Canada against the reciprocity treaty in 1911 helped to defeat him as a nominee of the Democratic party. [3] Woodrow Wilson was a native of Virginia, and his election, first as Governor of New

---

[1] Ibid., p. 644.
[2] Ibid., p. 638.
[3] C. B. Davis, "The Great American Novel——" (New York, 1938), p. 146. Josephus Daniels claimed that he would have won with the radio as Hoover did later. See James Kerney, The Political Education of Woodrow Wilson (New York, 1926).

Jersey, and then as President, pointed to a return of southern influence in the Democratic party. The long period in the wilderness was followed by aggressive legislation in the fields of both tariff and monetary reform. In Wilson's second term, begun with a narrow majority, patronage played an important role in maintaining the discipline of the party. After the outbreak of war, Wilson, according to Lindsay Rogers, became King, Prime Minister, Commander-in-Chief, party leader, economic dictator, and Secretary of State for Foreign Affairs. In the words of Josephus Daniels, "My party has the responsibility of this war." Exclusion of Republicans from the peace delegation meant that Wilson's promises became party politics.

The overwhelming burdens of the war on the executive took their toll in the breakdown of the President's health, in the defeat of the League of Nations by Congress, and in the nomination of Warren Harding from Ohio (1921-3), "the fine and perfect flower of the cowardice and imbecility of the Senatorial cabal that charged itself with the management of the Republican convention" (*New York Times*). [1] Colonel George Harvey had played an important role in the election of Wilson but the latter feared the possible charge of support by New York interests, especially J. P. Morgan and Co. That his fear was justified is evident in the fact that he was given the nomination by the Democratic party partly as a result of Bryan's attack on Champ Clark's reliance on New York support. The alienation of Harvey by Wilson was followed by his aggressive interest in the election of Harding and by his appointment as Ambassador to the Court of St. James. [2] Roosevelt had regarded settlement of the Irish question as "most essential to the furtherance of friendship between America and Britain" [3] and Harvey took an active part in establishing the Irish Free State and weakening support of the Irish vote to the Democratic party. He was instrumental in carrying out the views of the British and Americans in bringing to an end the Anglo-Japanese alliance by a four power treaty.

The death of Harding in office meant the elevation to the presidency of Coolidge from New Hampshire (1923-9). The religious issue

---

[1] Cited Agar, The United States, p. 675.
[2] See W. F. Johnson, George Harvey (Boston, 1929), pp. 286 ff.
[3] A conversation with Sir Joseph Ward, Prime Minister of New Zealand, in 1909. Hon. Sir James Kirwan, My Life's Adventure (London, 1936), p. 226.

was important in the defeat of Al. Smith [1] as it had been in the defeat of Seward [2] by Lincoln at the Republican convention in Chicago in 1860. The defeat of Hoover (1929-32) was in part a result of the jealousy of correspondents of the preferred position given to one of their number, Mark Sullivan, the difficulties of developing effective relations with the press in various administrative departments, and exploitation of this fact by Charles Michelson in a smear Hoover campaign. Libel laws were avoided by resort to the privileges of the *Congressional Record*. [3]

The disastrous results of the bitter aftermath of the Civil War shown as late as in the uncomfortable position of President Wilson and the attitude of the Republican party toward the peace treaty, were ultimately evident in the successive readjustments of the terms of peace, in the collapse of 1929, and the election of President F. D. Roosevelt, formerly Governor of New York. He exploited to the full the systematic efforts of Theodore Roosevelt to rid the name of association with the aristocracy. [4] Extensive control over patronage, the advantage of radio in appealing to the people over the head of Congress, and the disciplined support of labour enabled him to dominate the party until his death and enabled the party to dominate Congress to the present. "The radio ... the supreme test for a presidential candidate" was Roosevelt's "only means of full and free access to the people." [5] He was

---

[1] For a striking account of the implications of the Coolidge statement "I do not choose to run" for the final disposition of the Sacco Vanzetti case see We Saw It Happen ed. H. W. Baldwin and Shepard Stone (New York, 1938).
[2] Seward had been elected Governor in New York in 1838 with the support of the Roman Catholic Archbishop Hughes and had urged a division of the school fund between Catholics and Protestants with the result that he antagonized the strong American native party in Pennsylvania. McClure, Recollections of Half a Century, p. 216.
[3] Pollard, The Presidents and the Press, pp. 743-5.
[4] E. C. Bentley, Those Days (London, 1940), p. 198.
[5] R. E. Sherwood, Roosevelt and Hopkins (New York, 1950), pp. 184, 186-7. Every word in his speeches was judged not by appearance in print but by effectiveness over the radio and careful attention was given to accurate timing in relation to the number of words and the rate of delivery (pp. 217, 297). It is significant that before the radio no pre-eminent orator ever succeeded in reaching the presidency. A. K. McClure, Our Presidents and How We Make Them (New York, 1900), p. 88. It might also be noted that Blaine and Tilden were the only men who managed their own campaigns for the presidency and that both were defeated. Ibid., p. 312.

extremely sensitive to public opinion especially the opinion of religious groups. [1] The picture changed from one of a little-regarded presidential office and a supreme legislative branch under Harding, Coolidge, and Hoover and the strong position of business interests represented by lobbies, to one featuring a strong executive and a vast patronage to executive agencies. [2] In 1938 enormous relief funds were shifted toward preparation of armaments. [3] Even the Supreme Court which, as Chief Justice Hughes remarked, says what the Constitution is, generally sympathetic to the legislative branch of government, after a bitter struggle [4] became more sympathetic to the executive. Finally the transfer of the Bureau of the Budget from the Treasury Department gave the President access to all activities of the government.

The disequilibrium created by a press protected by the Bill of Rights had its effects in the Spanish American War, in the development of trial by newspaper, and in the hysteria after the First World War. Holmes wrote "when twenty years ago a vague tremor went over the earth and the word socialism began to be heard, I thought and I still think that fear was translated into doctrines that had no proper place in the Constitution or the common law." The effects of this hysteria were registered in the influence of the press on legislatures and on the Supreme Court (notable dissents only prove its strength). As a result power shifted increasingly to the executive and involved reliance of the executive on force. In the words of Brooks Adams: "Democracy in America has conspicuously, and decisively failed in the collective administration of common public property."

The power of the President in his control over patronage and party was not only enhanced by the radio but also by military considerations. The importance of the military factor strengthened the possibilities of leadership by a single person with power to intervene in war in spite of public opinion and of Congress. He was compelled to exercise wide discretion to lead or to force Congress to recognize and to accept his power and position. The position of the Democratic party and the President in the First World War, and in the Second

---

[1] Sherwood, Roosevelt and Hopkins, p. 384.
[2] R. G. Tugwell, "The New Deal: The Decline of Government," Western Political Quarterly, June 1951, pp. 295-312. For a study of the conflict between presidential and congressional authority over the administration see C. S. Hyneman, Bureaucracy in a Democracy (New York, 1950).
[3] Sherwood, Roosevelt and Hopkins, p. 101.
[4] J. Alsop and T. Catledge, The 168 Days (New York, 1938).

THE MILITARY IMPLICATIONS OF THE AMERICAN CONSTITUTION

World War, particularly as a result of the radio which widened the gap between the executive and the legislative branches, made it necessary to rely on important intermediaries—House in the case of President Wilson and Hopkins in the case of President F. D. Roosevelt. [1] In Great Britain by way of contrast the Prime Minister had the support of coalition and of Parliament. The solidity of the parliamentary tradition made it possible to defeat and to re-elect Churchill whereas the continued dominance of the Democratic party, while facilitating the transfer of power from Roosevelt to Truman, meant that changes could only be made in personnel, including members of the Cabinet. Americans were amazed at the necessity of Churchill's maintaining constant touch with the British Cabinet in drawing up the Atlantic Charter in Newfoundland in contrast with the independence of Roosevelt.

In the conduct of foreign affairs, a lack of continuity, [2] incidental to the importance of individuals, and in spite of the encouragement given to careermen in the Rogers Act of 1924, [3] was in strong contrast with the continuity evident in Great Britain and in Russia. This made for less attention to Europe, especially since the importance of interests in Latin America meant greater concern with ministers from these countries, particularly as they were men of ability and industry. [4] Difficulties in conducting negotiations with English representatives were evident at Bretton Woods, Washington and Savannah. English negotiators were constantly faced by Americans with the statement that they could not get that through Congress. The judgment of American negotiators as to the political tolerance of Congress and of public opinion became a determining consideration.

---

[1] Sherwood, Roosevelt and Hopkins, pp. 931-3. Ickes complained in 1940 that Hopkins had "never even attended a county meeting and wouldn't know how to get into one. Now here he is taking over a national convention. It's disgraceful." J. A. Farley, Jim Farley's Story: The Roosevelt Years (New York, 1948), p. 297.
[2] The diplomatic corps was an adjunct of the spoils system and the football of politicians. See Whelpley, American Public Opinion, pp. 113, 121.
[3] See Drew Pearson and R. S. Allen, Washington Merry-Go-Round (New York, 1931), p. 140.
[4] Ibid., pp. 30, 46.

III

The conflict between Cavalier and Roundhead, between absolute monarchy and absolute parliament, in England was transferred to North America. The southern colonies established at an earlier date reflected the influence of aristocratic organization and the northern colonies the influence of Puritan organization. The demands of the northern colonies for independence with relation to trade were paralleled by demands of the southern colonies for independence in relation to land. In the Revolutionary War the experience of George Washington in the colonial wars with the French became the basis for his appointment as military leader and in turn as President for two terms. He was followed by John Adams, a representative of New England, for one term. From 1801 to 1825 the three Presidents, Jefferson, Madison, and Monroe, each with two terms, were natives of Virginia. John Quincy Adams from New England served for one term and Andrew Jackson, a native of South Carolina, for two terms. He was followed in 1837 by Van Buren of the same party, the first President to be chosen from the middle colonies, who served one term. By this time the middle and northern colonies had built up the Whig party and succeeded by emphasizing military prestige in securing the election of General W. H. Harrison, followed by Tyler, a native of Virginia. The latter was followed by J. K. Polk, a native of North Carolina, and nominee of the Democrats. The Whigs nominated another military hero, while his laurels were still green, General Taylor, a native of Virginia, and again secured his election. In 1852 and in 1856 the Democrats succeeded by nominating weak northern Presidents, Pierce and Buchanan. Before 1861 all but two of fifteen administrations represented the Democratic party and of the thirteen nine were served by southern presidents. The Jefferson revolution from 1800 to 1860 was followed by Republican policy from 1860 to 1932.[1]

---

[1] McClure, Our Presidents and How We Make Them, p. 21.

THE MILITARY IMPLICATIONS OF THE AMERICAN CONSTITUTION

The dominance of representation from the South and especially Virginia, and of representation from the Army in the period prior to the Civil War, was a reflection of the dynamic power of the plantation system and its demand for more and better land. The weakness of the Spanish, Indians, and Mexicans made it possible for an aggressive government to steadily expand its territory to the west. Expansion of territory to the southwest gave an impetus to parallel expansion to the northwest to be accomplished with an occasional extension of territory at the expense of the British, for example in Maine and Oregon, and at the expense of the Russians on the north Pacific coast. In the race for land to the west and with its disappearance, the South attempted to expand territory for the slave trade along the northern border of the southern states. The friction eventually led to the outbreak of civil war or the war between the states.

With the end of the Civil War presidents were elected from the North and were again largely representative of the successful northern army. The aggressiveness of the North was checked by growing nationalism in Canada evident in controversies, over the fisheries centring around the Washington Treaty, the Alaska boundary dispute, and the reciprocity treaty of 1911. It took new forms in a continuation of the war against Spain and was effective in the addition of new territory.

Broadening of the powers of the executive such as those boasted about by Theodore Roosevelt and the improvement of communication notably in radio strengthened the position of the President. Control over vast sums following the depression and continued during the war enabled the President to control the party. The seven principles of politics, five loaves and two fishes, were handled more effectively. Patronage and assistance in elections were distributed in accordance with the record of the roll calls in Congress. [1] In the election of presidents directly by majority vote was registered the importance of the middle class urban vote, especially of New York, and the election of senators, following the abolition of election by caucus, [2] two from each state representing predominantly a rural middle class, increased

---

[1] George Michael, Handout (New York, 1935), p. 73.
[2] For a criticism of the direct primary see C. J. Stackpole, Behind the Scenes with a Newspaperman: Fifty Years in the Life of an Editor (Philadelphia, 1927).

the possibilities of friction. [1] The House of Representatives also reflected the influence of the urban vote but its size left it exposed to vicious partisan and predatory interests and to manipulation under stupid rules such as prevailed under Cannon and after 1925 under the Longworth Snell Tilson triumvirate. [2] It has been described as the greatest organized inferiority complex in the world.

With the tendency toward increased power in the executive and the increasing importance of urban centres the policy of parties is less dependent on a single figure in the presidency. Family names will probably persist as a factor in the selection of presidents—to mention Harrison, Roosevelt and Taft—and the dangers of assassination [3] will be checked by strengthening of the secret service. Formerly vice-presidents were selected as representatives of a defeated minority within the party and were consequently in a weak position when they rose to the presidency. [4] From 1800 to 1900 only one vice-president, Van Buren, was elected in his own right to the presidency. [5] More recently the Vice-President has become a regional representative intended to support the President as a representative of a densely populated state. Garner from Texas supplemented Roosevelt as did Wallace from Iowa and Truman from Missouri. Since 1900 three Vice-Presidents have been elected in their own right: Roosevelt, Coolidge, and Truman.

The importance of New York State and of the possibilities of rapid advance in political life by attacks on corruption explained the prominence of Tilden who attacked the Tweed ring and as Democratic candidate opposed Hayes; of Cleveland who made his reputation in Buffalo; of T. R. Roosevelt, who was New York police commissioner; of Charles Hughes, Republican candidate in opposition to Wilson, who came into prominence in the insurance investigation; and of Dewey

---

[1] A. N. Holcombe, The Middle Classes in American Politics (Cambridge, Mass., 1940), p. 104.
[2] Pearson and Allen, Washington Merry-Go-Round, pp. 217-19.
[3] The influence of anarchism and the Colt revolver on the disappearance of apparent dictatorships in business and in governments has never been given careful study. See Emma Goldman, Living My Life (New York, 1934).
[4] H. L. Stoddard, As I Knew Them: Presidents and Politicians from Grant to Coolidge (New York, 1927), p. 123.
[5] McClure, Our Presidents, p. 25.

with his prosecutions. The intensity of the struggle in New York [1] was evident in the efforts of Hearst to become mayor and governor and eventually president. Coolidge emerged as a national figure in the Boston police strike. Perhaps the comparatively healthy state of New York in spite of the scale of its problems has been partly a result of its possibilities in the making of reputations by attacks on corruption.

The President cognizant of his power must be constantly alert to the implication of policy for voting strength. In foreign policy the results have been evident in several directions. Timing has been carefully worked out in relation to voting or rather voting has been carefully planned in relation to time. A rigid time arrangement compels an emphasis on manoeuvrability or the settlement of issues when the effects will be most evident in relation to votes. Mr. Truman immediately before the election in 1948 decided to recognize Palestine and to strengthen the position of the Democratic party in New York State of which Mr. Dewey was Governor. A period of tension and war enormously increases the executive power. The opposition is prevented on the large vague grounds of security and military secrecy from discussing effectively the most crucial element of policy. During the war Republicans were appointed to the Cabinet and bi-partisan responsibility in foreign affairs was assumed. The argument about swapping horses in midstream has proved difficult to answer. It might be answered by nominating a general, let us say Eisenhower, but West Point has never produced good politicians, and he may be content with actually having more power than the President. F. D. Roosevelt, with a personal interest in the Navy, left Army experts with much greater freedom of decision. [2] Such freedom, however, tends to throw the President into the hands of the armed forces. The two-thirds rule regarding treaties in the Senate has been effective in checking the foreign policy of presidents and has been exploited by German, Rus-

---

[1] The Democratic party in New York State became a political workshop of the United States and leaders throughout the United States after 1925 were urged to organize along the lines of New York, especially in giving women an equal voice on committees. The feminine vote became an important factor in 1932 and in 1936. See James A. Farley, Behind the Ballots: The Personal History of a Politician (New York, 1938), pp. 55, 160.

[2] Churchill exercised much greater control over the army. See Sherwood, Roosevelt and Hopkins, p. 246.

sian, and Clan-na-Gail delegations, [1] but it has been of little avail with the development of the United Nations and the power of armed force. Indeed the Senate has shown considerable readiness at the demands of the party to co-operate with armed forces.

In the twentieth century the enormous development of industry accentuated by war has greatly enhanced the problems of the executive. Use of the blockade and the threat of blockade has increased dependence on domestic industries. "An all-round increase in armed forces" has been necessary "to mitigate unemployment." We must have "war to solve unemployment in order to ensure against internal anarchy, instead of war solely to protect employment (ordered life) against external aggression." "The dependence on war has become even more vital to our economic system than the dependence of war on industry." "Should an enemy not exist he will have to be created." [2] "A war cannot be carried on without atrocity stories for the home market." [3]

---

[1] Count Cassini, a Russian minister, and Von Holheben, a German minister, appealed successfully through the press to the Senate against presidential policy. The Education of Henry Adams, p. 375.
[2] Fuller, Armament and History, pp. 164-5.
[3] Bentley, Those Days, p. 184.

IV

These remarks have been made by one who does not pretend to understand the United States and who cannot appraise the significance of the party struggle as part of the domestic scene. But we are required in the interests of peace to make every effort to understand the effects not only of the actions of the United States but also of our own actions. We have never had the courage of Yugoslavia in relation to Russia and we have never produced a Tito. We have responded to the demands of the United States sometimes with enthusiasm and sometimes under protest. Members of the British Commonwealth struck back against the Hawley-Smoot tariff in the Ottawa Agreements. But we have been a part of the North American continent. The enormous increase in the production of wheat on this continent in the last century was directly related to the Russian revolution, the rise of agrarianism in Germany, of higher tariffs in France and of marked adjustments in England. Germany imposed a tariff on sugar to secure independence in supplies of sugar, drove down the prices of cane sugar, contributed to the outbreak of revolt in the Spanish American colonies, and enabled the United States to take full advantage of the break-up of the Spanish empire.[1] The immigration quota of American legislation in 1924 accentuated the population problems of Italy and contributed to fascism. The silver purchase agreement of 1934 and the consequent destruction of the Chinese monetary system were related to the revolution in China. The protectionist policy of North America and the difficulties of penetrating the American market compel the United States to export dollars and at the same time make it difficult for other countries to acquire dollars. As a result there is resort to enormous expenditure on armament. In the words of the late Carl Becker, what we didn't know hurt us a lot.

---

[1] See Brooks Adams, America's Economic Supremacy (New York, 1900), pp. 36-41.

A written constitution with its divisive nature established by the Declaration of Independence and the Constitution, centralization under Washington and Adams, decentralization from Jefferson to Lincoln, and centralization after Lincoln, first under the Republican party and later the Democratic party, so that at one time there has been a weakening of the power of the executive and at another a strengthening of that power depending largely on the dominant medium of communication, stand in sharp contrast with the unwritten constitution of Great Britain and the undivided power of the Prime Minister responsible to Parliament. In the United States parties are "devoted to the search for compromise between sectional, class, and business groups" and are "frankly uninterested in logical programs or 'eternal' principles." [1] The practice of representation from party rather than regions characteristic of Great Britain finds no expression in the United States. [2] "The most profound of American political thinkers saw in the perpetual search for compromise between selfish interests the basic principle of free government." In the words of Calhoun, "the negative power ... makes the constitution,—and the positive ... makes the government. The one is the power of acting;—and the other the power of preventing or arresting action. The two, combined, make constitutional government." [3] The emphasis on negation, the constant fear of Leviathan, of the encroaching state, has been offset by the promotion of strong government by war and industrial revolution. [4] Under the American Constitution reliance on force has become increasingly necessary whereas under the British, following the brief period in which Parliament was dominated by Cromwell and the army and the period in which the Duke of Wellington was Prime Minister, force has been increasingly subjected to the authority of Parliament. A general as Prime Minister of England would be unthinkable, though the influence of the

---

[1] Agar, The United States, p. vii.
[2] The fathers were particularly concerned to avoid the borough system of England. "State law and custom have practically established that a representative must be a resident of the district from which he is elected." See D. A. S. Alexander, History and Procedure of the House of Representatives (Boston, 1916), p. 5. As a result the mobility of the ablest individuals has been checked, whereas in England parties have been much more effective in attracting and securing the election of the ablest individuals irrespective of residence.
[3] Agar, The United States, p. vii.
[4] Ibid., p. xiii.

army and navy are not to be disregarded, whereas in the United States a general as President has been regarded almost as a rule. Ostrogorski has quoted the remark that God looks after little children, drunken men, and the United States. I hope it will not be thought blasphemous if I express the wish that He take an occasional glance in the direction of the rest of us.

ROMAN LAW AND THE BRITISH EMPIRE

By

HAROLD A. INNIS

Dean of Graduate Studies and Professor of Political Economy
University of Toronto

*One of a series of lectures commemorating
the 150th anniversary of the University*

Delivered at THE UNIVERSITY OF NEW BRUNSWICK
MARCH 30, 1950

When your President asked me to participate in a programme to celebrate the hundred and fiftieth anniversary of the University of New Brunswick I found it impossible to refuse, since it is an institution particularly close to my heart as the first to give me an honorary degree and since your President is an old and persuasive friend. I shall not in this gathering, where his reputation stands so high, describe his methods of persuasion, not that I shall soon forget them. I have perhaps a further reason since New Brunswick is an ancestral home from which my forbears with others moved to Upper Canada.

With characteristic generosity under the circumstances, the President has given me complete freedom in the selection of a subject. It seemed fitting that I should be concerned with a country which has played an important role in the life of this institution, namely the United States. This province was created as a result of strategic plans of defense on the part of the second British empire against the colonies which had rebelled. Nova Scotia was divided into three separate areas, Cape Breton, New Brunswick and Nova Scotia in order to provide separate nuclei around which defensive measures might be mobilized. Loyalists migrated to New Brunswick and kept alive the memories of hostility to their native land. Christopher Sauer, a prominent figure in the history of printing in Pennsylvania, started the first newspaper in New Brunswick. This university began as your calendar states through the interest of loyalists in the education of their children and, in the words of the memorial in 1785, the "necessity and expediency of an early attention to the establishment in this infant province of an academy of liberal arts and sciences." It would be ungracious of me to elaborate on the contribution of universities to the Maritime provinces and to western civilization, since that has been done so ably and so fittingly by the late Sir Robert Falconer, the late J. C. Webster, Professor D. C. Harvey and your own professor A. G. Bailey, Mrs. C. P. Wright and others.

The late James Bryce attempted to throw light on the problems of the British Empire by emphasizing parallels with the Roman Empire and in particular suggesting the contributions of Roman law and of common law to the development of the respective empires.[1] At the very period in which Bryce was revising his essays for separate publication in 1914 the British Empire was undergoing crucial change. Since they were published the development of the Commonwealth after the first world war in the Statute of Westminster and the changes in status of Ireland, India and Newfoundland point to a need for reconsideration.

The name of Bryce will always be associated with the results of the first major change in the Empire in *The American Commonwealth*. The American Revolution was a result of limitations of common law which have been discussed by a large number of English, American and other scholars. Prof. C. H. McIlwain has described the problem of common law in the seventeenth century when parliament reflected the influence of force in the substitution of the Cromwellian regime for that of the

---

[1] James Bryce, The Ancient Roman Empire and the British Empire in India: The Diffusion of Roman and English Law Throughout the World; Two Historical Studies (London, 1914).

Stuarts. The absolute power of the Tudors was replaced by the absolute power of parliament and both were regarded as encroachments on common law. Sir Edward Coke defended the position of common law as stated in the Bonham case in 1610. "When an act of Parliament is against common right and reason, or repugnant, or impossible to be performed, the common law will control it, and adjudge such act to be void." But such limitations were not recognized by parliament under Cromwell or in the establishment of legal supremacy in the Revolution of 1689. The English colonies in North America had been established in the period before parliament had assumed this position and were unable to accept its implications. James Otis restated the position of Coke, and the Assembly of Massachusetts on March 2, 1773, refused to recognize the supremacy of Parliament. "We conceive that upon the feudal principles all power is in the king; they afford us no idea of parliament." Great Britain had seen the evolution of the supremacy of parliament at the expense of common law, and the colonies, determined to protect the position of common law, introduced a constitution designed to check the power of legislative machinery.

It will not be necessary to rehearse the steps taken by Great Britain, and the colonies remaining within the empire, to develop a constitution which would evade the disaster of the first empire. The Maritime provinces succeeded in building a second empire from the wreckage of the first in which responsible government was achieved. The common law came into its own with a recognition in Great Britain of the limitations of parliament and recognition in the colonies that the elaborate machinery of the United States to protect the common law was unnecessary. In Great Britain the effects of the common law were evident in the Reform bills and in the extension of the franchise in the 19th century. Elements in the constitution opposed to its effective operation were steadily weakened as the House of Commons increased in power at the expense of the House of Lords. Long and bitter struggles characterized the change and still characterize it but the legislation of 1911 definitely brought the power of the Lords to an end. "The House of Commons after putting under its feet the Crown and the House of Lords, has in its turn been put under the feet of the caucus." [1]

The changes within Great Britain had profound implications for the Empire. Indeed the legislation of 1911 was directly linked to the problem of Ireland and the possibility of establishing Home Rule. Defeat of

---

[1] Goldwin Smith, Essays on Questions of the Day (New York, 1893) p. 98.

the Conservative party was followed by opposition to the Liberal party supported by Irish and labor members first in the House of Lords and finally in Ulster. The unsavoury story in which the army joined hands with Ulster leaders and leaders of the Conservative party, described by Prof. M. J. Bonn [1] as the beginnings of fascism in Europe, need not be retold at this point. For the first time parliament was openly and to some extent successfully defied by force. During the first world war Irish opposition became more determined and led to the Easter rising of 1916 and eventually to the treaty and the Irish republic. A common law parliament had become impossible in the face of obstructionist tactics which developed from the Irish question.

It has been suggested [2] that British imperialism succeeded in areas in which native populations were eliminated as in America and Australia or in areas in which a bureaucracy could be established as in India and failed in areas in which a strong cultural influence dominated garrisons of settlers as in Ireland, but the suggestion overlooks the role of common law. Men trained in common law such as Gandhi were quick to see its possibilities in the protection of the rights of individuals. After his training in London, Gandhi carried on an effective campaign in South Africa on behalf of Indian immigrants, and with the techniques developed in South Africa contributed powerfully to the establishment of India and Pakistan. A common law basis implied concern with local customs and facilitated the development of the British Commonwealth by peaceful means or by minor rebellions.

We have perhaps said sufficient to indicate that the British Empire under the influence of common law has pursued a vastly different course from that of the Roman Empire. We may now inquire more directly into the characteristics of common law. Various writers have discussed the origins of common law in England to show that it consisted of customs which existed in unwritten form and that it was necessary to discover these customs through the use of the jury system and the calling together of representatives of different communities in parliament. The words writ, oath, witness and possibly gallows did not originate in France. Parliament was concerned with the protection of individuals and not with the provision of privileges enabling members to abuse individuals outside its walls. In the words of Pollard, "A

---

[1] M. J. Bonn, Wandering Scholar (New York, 1948) p. 89; see also George Dangerfield, The Strange Death of Liberal England (New York, 1935).
[2] M. J. Bonn, op. cit., p. 101.

foundation of common law was indispensable to a house of common politics." Parliament until the rebellion of the seventeenth century was pre-eminently judicial rather than legislative. [1] With the increasing importance of legislation particularly after the reform act of 1832 lawyers continued to play an important role in parliament in the making and in the interpretation of statutes. Common law countries favour the election of lawyers as legislators to the exclusion for example of journalists, in contrast with Roman law countries which seem to favour journalists as legislators. In common law countries the state became a part of customs and traditions and the revolutionary tradition was weakened. Marx's withering of the state had reference to Roman law and not common law countries. Common law traditions which made politics a part of law and emphasized the relation of the state to law implied an absorption of energies in politics and a neglect of the cultural development which has characterized Roman law countries. The danger of imposing common law traditions on Roman law countries has been evident in the difficulties of the parliamentary system in those countries.

The implications of the dominance of lawyers are suggested in remarks by Sir Henry Taylor.

> "Of law-bred statesmen (if they have had practice at the bar) the peculiar merit is a more strenuous application of their minds to business than is often to be found in others. But they labour under no light counterpoise of peculiar demerit. It is a truth, though it may seem at first sight like a paradox, that in the affairs of life the reason may pervert the judgment. The straightforward view of things may be lost by considering them too closely and too curiously. When a naturally acute faculty of reasoning has had that high cultivation which the study and practice of the law affords, the wisdom of political as well as of common life will be to know how to lay it aside, and on proper occasions to arrive at conclusions by a grasp; substituting for a chain of arguments that almost unconscious process by which persons of strong natural understanding

---

[1] C. H. McIlwain, The High Court of Parliament and its Supremacy, an historical essay on the boundaries between legislation and adjudication in England (New Haven, 1934).

get right upon questions of common life, however in the art of reasoning unexercised.

The fault of a law-bred mind lies commonly in seeing too much of the question, not seeing its parts in their due proportions, and not knowing how much of material to throw overboard in order to bring a subject within the compass of human judgment. In large matters largely entertained, the symmetry and perspective in which they should be presented to the judgment requires that some considerations should be as if unseen by reason of their smallness and that some distant bearings should dwindle into nothing. A lawyer will frequently be found busy in much pinching of a case and no embracing of it—in routing and grunting and tearing up the soil to get at a grain of the subject;—in short, he will often aim at a degree of completeness and exactness which is excellent in itself, but altogether disproportionate to the dimensions of political affairs or at least to those of certain classes of them." [1]

As has been said of many lawyers all acts are to them free and equal. An elementary discussion of the conditions under which lawyers work in practising at the bar from a lay point of view may suggest more clearly the important role of the legal profession. The arrangement of the court room emphasizes power and authority such as characterize proceedings involving life and death. The bench sets off sharply the position of the judge, and below him the witness stand, the bar for opposing counsel, an inner bar for His Majesty's counsel learned in the law, and beyond seats for the public. The tradition of awe inspired by these arrangements, insistence on the dignity of the court and rigid prohibitions against smoking, chewing gum or other distractions such as may include the reading from a manuscript by counsel inspire a concern with the search for truth and justice.

Encroachment on these traditions has been evident in the demand for photographs for the press and in the interest of criminal lawyers in publicity. The court has possibilities of advertisement for young lawyers. Even members of the supreme court appear to relish the appearance of their photographs in the press. But while lawyers dis-

---

[1] Sir Henry Taylor, Notes from Life—The Statesman (London, 1878) p. 384-5.

play a keen interest in the details of crime such as those appearing in the press they tend to dislike specialization in criminal law and to prefer a mixed practice of civil and criminal law. Concentration on criminal law is apt to be thought of as having a deteriorating effect on character and reputation.

Procedure involves dependence on the oral tradition in eliciting testimony from witnesses who have been placed under oath to give the truth, the whole truth and nothing but the truth. Facts are determined by examination and cross-examination and re-examination of counsel. Opposition between counsel is designed to check and to produce evidence from which the judge or the jury must decide the case. When evidence has been elicited and established argument to establish the law suited to the facts follows. Respect will be shown in language and demeanour to the bench—cases of dispute with the bench always being prefaced by the words "with great respect." The maxim handed on to young members of the bar "never talk down to the bench" reflects the egoism of the bar and the necessity of emphasizing the place of the bench. The significance of the oral tradition is evident in the possibility of checking extravagant statements made by counsel or by witnesses. With a background of development prior to the spread of reading and writing, the tradition of the importance of oral rather than written evidence has persisted in the procedure of the court and in the jury system. The common law has consequently been responsive to the opinion of all classes of society including the illiterate. This contact has possibly been more effective than that of the church and religion since it is without the elaborate ceremonial and the written scriptures of the latter though it musters support from religion in requiring testimony sworn on the Bible and may exact severe penalties for perjury. English courts will insist on the appearance of living authorities rather than extracts from text-books written by them on the assumption that such an authority may have changed his mind after writing the book. In North America the difficulty of transporting a living authority over long distances has favoured a whittling down of the English rule and increasing reliance on the text. The advantage of the oral tradition shown in its sensitivity to constant change even during the course of the trial becomes evident in the exposure of weaknesses in evidence and in argument. The character of witnesses is brought out in detail and the role of intent more easily established. In the preparation of cases counsel must study intensively the character of his own and other witnesses and estimate strong and weak points if he is to work out satisfactory tactics in presentation. The common law gives great emphasis to character and to the study of character from an objective point of view. Its success is linked to individualism and necessitates a concern with the influence of the state on character and of character on the state. There is danger of forgetting the words of the Lord Chancellor (2 Eden. 113) "Necessitous men are not, truly speaking, free men."[1] "It is precisely because the force of circumstances always tends to destroy equality: that the force of legislation should always tend to maintain it." (Rousseau).

I am tempted to insert an editorial by Albany Fonblanque, written in England in the middle of the 19th century.

> "It was but the other day, however, that a most tender and touching sight was presented in Lord Carlisle's Court of Inquiry—Mr. Serjeant Wilkins weeping for Mr. Ramshay, his learned bewigged head bent to the table 'like a lily borne down by the hail'. Perhaps, prosaically, it was more like a cauliflower on a block, but let that pass. What we have to consider is the zeal, or the fee-compelling-force, which can bow a wigged head to the table, and make the eyes overflow with tears such as either genuine pity, or genuine onion, elicits—tears such as learned serjeants shed. The eye that so weeps, however, must have seen a fee. An unfeed eye would on a similar occasion be as unmoved as a stone. The fee and the feelings go together: the word feeling, in legal diction, being derived from fee. What the precise charge for weeping is we do not pretend to know; nor whether it is set down in the brief as an extra, like consultation, or a refresher: but of late years we have had several exhibitions of this black grace. Chitty wept for Thurtell, and Fitroy Kelly for Tawell, and lastly Wilkins for Ramshay. Sweet sensibility! says the tender-hearted reader; but how is it that this same sensibility of the learned is so capricious, and that the same wigged man, who blubbers over one client so affectingly, will throw another overboard without a hesitation or a scruple? Why make fish of one and flesh of another? Why so strain the duty of advocate and client in

---

[1] Cited E. S. Corwin, The Twilight of the Supreme Court (New Haven, 1934) p. 207.

some show cases, and loosen it in others, as we see in this example?

"The complaisant husband who had napped during Caesar's visits, on finding that the same somnolency was expected from him by another gallant said, 'I do not slumber for everybody'. Mr. Serjeant Wilkins does not sob for everybody: but in common fairness and honesty he is bound to explain the rules of his service or disservice to his clients, specifying for which of them he goes through thick and thin, and which he throws overboard." (1851) [1]

In stressing the importance of the oral tradition it is necessary to keep in mind the role of the written and the printed tradition. In England courts are more jealous of their position and check discussion by newspapers when cases are *sub judice*. The dangers of extravagant publicity become acute when members of the jury may have come under the influence of public opinion reflected in the press. A more subtle problem arises with the spread of mechanization in reports of proceedings of the court. Since questions and answers are phrased in relation to a sworn record which may become the basis of consideration and decision by the bench they will tend to blur the sharp impressions characteristic of an oral tradition. The oral tradition is carefully warped in relation to the demands of a written or stenographic tradition. A concern with the record implies an interest in a type of question suited to reading and a neglect of the transient impression of the spoken word. The tendency to concentrate on the record has an advantage in that it enables the bench to study the case in a dispassionate and objective fashion but a disadvantage in that it enables the bench to delay reaching a decision and perhaps encourages continuance on the bench of men who by age or inclination are reluctant to appreciate the importance of promptness in the administration of justice. But there may be warrant for the remark that truth will out even in an affidavit. The legal profession in itself has an important influence on the administration of justice. Counsel are constantly alert to the artistic character of work done by members of the profession and are continually engaged in the appraisal of respective capacities of fellow members and of those of their ranks appointed to the bench. The es-

---

[1] The Life and Labours of Albany Fonblanque ed. by his nephew, E. B. de Fonblanque (London, 1874) p. 340-1.

sentially feudal character of the legal profession is evident in references to my lord, my friend or my learned friend. Style has become more prosaic and matter of fact and even conveyancing can perhaps no longer be described as "a jungle of antiquated fooleries kept up by the pedantry and interest of those who profited by it." [1]

The clashes between opposing counsel bring out sharply the competition in ability. Each interest appearing before the court is obsessed with its own advancement and becomes extremely critical of counsel in cases of defeat. Courtesies between members of the legal profession temper the acerbities of conflict between interests and impose a severe restraint on bitterness. The appearance of conflict in the courts will meet the demands of interested parties and permit the courtesies of the profession outside the court. The protection of the courts and the interest of counsel in clients ensures that questions of fact of an embarrassing character will be brought out but the relative capacities of counsel are apt to be reflected in the size of fees paid to counsel and in the ability of interests to pay fees. Success will depend on the ability of counsel but also on the size of the legal firm. A large firm acquires enormous resources in the specialized knowledge of its members and its ability to attract energetic and able young juniors. The demands for intense industry can only be met by younger men and explain the general impression of relatively short lives in the profession. The advantage of the large firm has become more evident in the enormous increase in legislation and in the numbers of digests, indexes and abridgements of reports of cases. The large amount of printed material has been further increased by the production of text books, commentaries and the like and the growth of black letter law. Lawyers tend to become lazy with the increase of indexes and digests, to neglect a reading of cases with thoroughness and system and to demand more indexes. Large earnings assume an enormous importance in the administration of law. Ability is maintained in the bar and restricted on the bench which is apt to be impressed by counsel capable of securing large fees. There appears to be a tendency for large companies to secure protection in legal counsel and for counsel to be able to win large fees in successfully protecting them. Law tends to favour those earning large fees and to have a commercialistic bent especially with need for an expensive library and the use of abridgements. The spread of printing weakens the oral tradition. The increasing importance of the

---

[1] Frederic Harrison, Autobiographic Memories (London, 1911) I, p. 149.

written mimeographed and printed tradition has been accompanied by a decline in the position of the courts and changes in the character of law.[1]

Executorships of wills have largely gone to trust companies and account collections to collection agencies. A marked increase in the mortgage business of insurance and loan companies has led to specialization and the handling of business by larger law firms. So too corporation work has become highly specialized and has come into the hands of large firms. Practice of law in relation to automobile accidents has fallen into the hands of lawyers acting for insurance companies. Income tax law has become the concern of legal specialists who are forced to compete with chartered accountants. Labor law has become a special field. The rise of boards of an administrative character has meant a demand for specialists other than lawyers. Law has followed the shift from individualism to collectivism. Able young graduates from law schools are apt to become immediately interested in office work rather than court work to the great disadvantage of courts. Demands on the legal profession have increased with the specialization which characterizes the Western World. Cases are presented before modern courts involving a mastery of highly technical questions in a wide range of subjects. The expert appearing as a witness, whether accountant, economist, engineer or doctor must be subjected to intelligent examination and cross-examination involving a mastery on the part of counsel of the particular subject under consideration. There is a well known maxim to the effect that one should never ask a question in cross-examination of which one does not know the answer. The legal profession must maintain a profound belief in its capacity to master any evidence and to adapt all questions to the demands of the court. Counsel are compelled to concentrate intensively on particular problems and to become obsessed with a knowledge of immediate details. The common law with its emphasis on the oral tradition has perhaps a greater interest in the ascertainment of facts than other legal systems. Facts are more important than principles. Litigious procedure for example emphasizes circumstantial evidence in contrast with the inquisitorial procedure of code countries. The importance of the jury system and opposition to the use of hearsay evidence through the fear of misinterpretation by the jury stands in contrast with other

---

[1] I am indebted to Mr. F. M. Covert, K. C. and Mr. G. Demarais, K. C., for general views on this subject.

systems and involves its own handicaps. For example I am told that a purchase from a department store can be proved by an appeal to the sales clerk but not by reference to the more certain evidence of the department store's records. The advantages of the common law system with its emphasis on facts are probably evident in a society favorable to the scientific tradition and industrial development in the sense developed by Bacon. It is further evident in the emphasis of a common law society on news. Lawyers reflect the interests of newspapers in questions of the moment. These advantages assume limitations. Considerations involving continuity in time are rather neglected and the long term factors ignored. A training in law makes for a brittle, brilliant type of work. Lawyers are compelled to master the intricacies of a case and after its completion to forget it and to master the intricacies of the next case. The memory tends to be neglected, general principles to have limited attraction, and general theory to be ignored. Law is apt to become anything "boldly asserted and plausibly maintained". A neglect of the time problem implies a lack of interest in theoretical problems. In contrast the Roman law tradition in its concern with principles attracts the highest intellectual ability to the academic field and enhances an interest in philosophical theory and theoretical speculation. In turn it becomes possible to develop an interest in problems of continuity of time, though the late Justice Holmes could write "People want to know under what circumstances, and how far, they will run the risk of coming against what is so much stronger than themselves, and hence it becomes a business to find out when this danger is to be feared. The object of our study then is prediction of the incidence of the public force through the instrumentality of the courts.... For the most important and pretty nearly the whole meaning of every new effort of legal thought is to make these prophecies more precise, and to generalize them into a thoroughly connected system." [1]

The capacity of large fees to attract able counsel weakens the possibility of attracting them to the bench or to political life but the bench has become attractive as a result of income tax regulations and a prospect of holidays. It has been pointed out that separation of the barrister and the solicitor in England tempers the effect of finance on the legal profession and that the combination of the two positions in the solicitor in Canada greatly enhances the impact of business and finance on the legal profession. During periods of depression with de-

---

[1] Max Lerner, The Mind and Faith of Justice Holmes (Boston, 1943) p. 72.

cline in fees counsel will perhaps turn more quickly to political activity. Reluctance to forego large fees in the large cities tends to favour acceptance of appointments to the provincial bench rather than the Supreme Court in Ottawa. Dislike of living in Ottawa is accompanied by the prestige of provincial supreme courts in provincial capitals. Relative absence of restrictions on age of retirement on the provincial bench as compared with the federal court enhances the attractions of provincial courts and explains to an important extent the relatively high caliber of provincial appointments. Since the salaries of judges in the provinces are uniform, appointments in the smaller provinces with lower living costs and much less business become extremely attractive. Consequently lawyers assume an intense interest in politics and premiers have become chief justices of the provinces. Politics are apt to be dominated by lawyers and to be slanted in the interest of lawyers. Appointments to the federal Supreme Court and to the provincial courts on the other hand are subject to restrictions in religion, region and language. The Province of Quebec, partly because of the importance of the civil code as well as common law, because of French and English, has been given three judges, and in turn the Province of Ontario is represented by the same number one of whom must be an Irish Catholic. The Maritimes are represented by one member and the Western provinces by two members. The rigidity of conventions in appointments reflects the power of the legal profession to defend its interests. The domination of the Liberal Party in the House of Commons, the Senate and the judiciary assumes a monopoly of legal knowledge. The effects of these restrictions will be tested more sharply with the abolition of appeals to the Privy Council and they may well prove to have serious consequences for the success of the federal system of government.

    Reluctance to accept appointments on the bench because of the attraction of large fees tends to divide the profession into two groups. Appointments to the bench are essentially political and counsel less attracted to the court are compelled to recognize the importance of political activity. The second group of lawyers therefore enter parliament and have a direct effect on legislation through statutes and following a political career receive appointments to the bench before whom practising lawyers must appear. Successful practising lawyers are compelled to interpret legislation prepared by and to practise before successful political lawyers. Counsel trained in the common law tradition in parliament and on the bench are concerned with legislation reflecting a common denominator of public opinion and registering the effects of a training with an emphasis on facts. Legal training which assumes a capacity to ascertain and to master factual presentation ensures that parliament has at its command an array of ability particularly adapted to its varied demands in the enactment of legislation covering a wide variety of subjects. The effects of legal training shown in the capacity for intense concentration and the mastery of facts in a short period of time have been evident in the success of lawyers in political life. The effectiveness of legal Prime Ministers can be illustrated with reference to Lloyd George not to mention illustrations nearer home. In the words of Lloyd George "I should always feel at liberty to override the findings of any body of experts." [1] Lack of pensions for politicians contributes to the attraction of parliament to lawyers whose chances of appointment to the bench have been greatly improved by political activity. The hazards of political life for the lay politician and the absence of political pensions accentuates the competition among lawyers for the bench or for the Senate. As has been said of the United States Supreme Court "The court is small, the cream (sometimes not very fat cream) of a profession in which the political impulse is strong." [2]

Traditions of procedure emphasizing the oral tradition in common law countries in the court and in parliament imply a background unsympathetic to the social sciences with their emphasis on the written tradition. Inclusion of courses in the social sciences in the lawyer's and of courses in law in the training of the social scientist may contribute to a solution of the difficulty and to a reconciliation between law and the social sciences but on the other hand may weaken the distinctive contribution of each. The advantages in a legal training which permits a rapid shift from the intricacies of one case to those of another are offset by an inability to penetrate problems to an appreciable depth, whereas the advantages of a training in the social sciences in the mastery of complex problems are offset by an inability to shift quickly from the intricacies of one problem to another. The long and tedious process of working through complex problems of the social sciences is in sharp contrast with the demand for swift effective argument in the law courts. Cross fertilization quickly reaches a point at which its advantages are followed by the disadvantages of cross sterilization. The type of social scientists acceptable to the courts is marked by the abil-

---

[1] Valentine Williams, The World of Action (Cambridge, 1938) p. 309.
[2] E. S. Corwin, op. cit., p. 54.

ity to ask questions intelligible to lawyers and to answer questions intelligible to lawyers. This type of social scientist rarely enhances his prestige among his fellow social scientists and appears eventually to lose his prestige even among lawyers who in turn become contemptuous of the complications of the social sciences. Social scientists concerned with fine spun abstractions tend to neglect a sense of proportion and the practical matters of fact with which common lawyers are obsessed. Social scientists appearing in common law courts are necessarily concerned with immediate problems and are consequently restricted in the development and application of theory. They tend to become advocates and to reflect the points of view of their employers. The longest purse will produce the best economist. The late Justice Holmes may have been right in saying that "for the rational study of the law, the black letter man may be the man of the present; but the man of the future is the man of statistics and the master of economics" [1] and that "every lawyer ought to seek an understanding of economics" but he was certainly accurate when he said that "the present divorce between the schools of political economy and law seems to me an evidence of how much progress in philosophical study still remains to be made." [2] It is the function of the social sciences and the bureaucracies to offset the effects of the obsession of common law with nominalism. The hierarchy of the law undoubtedly weakened the ecclesiastical and military hierarchies. It has been influential in the development of an effective business hierarchy which has dangers for the hierarchy of law itself. The place of lawyers in business is strengthened by their status in the courts and the place of lawyers in the courts is strengthened by their status in business.

Following these remarks on the character and implications of common law I propose to turn to a discussion of the influence of Roman law in the British Empire. The British emerged in part as a result of a balance between the oral tradition and the written tradition, between common law and Roman law. [3] The element of Roman law, especially as reflected in the canon law, which persisted after the Reformation in England was gradually reduced in importance in the British Empire and results were evident in the Commonwealth. The divine right of the papacy was replaced by the divine right of Parlia-

---

[1] Max Lerner, op. cit., p. 83.
[2] Ibid., pp. 85-6.
[3] See F. W. Maitland, English Law and the Renaissance (Cambridge. 1901).

ment after the rebellion. Following the submergence of the concept of fundamental law which eventually precipitated the American Revolution, the written constitution of the United States was designed to restore it and to protect its position. Emergence of a federal government in a constitution which gave enormous powers to the courts involved protection to fundamental law but in protest against the divine right of parliament assumed the divine right of the United States. Without a written constitution Great Britain was able eventually to master the problem of Empire and to digest the element of Roman law or rather to cast it out into regions which left the Empire as in the United States or regions which insisted on independence and autonomy within the Empire as in members of the Commonwealth.

The element of Roman law which became more powerful in other parts of the Empire was evident in the insistence of small areas on their autonomy and divine rights, [1] in the emergence of a federal system and in conflicts over the concept ending in the United States in the war between the states. Temporarily its significance was lessened but supremacy in the north reflected the importance of the divine right of union essential to effective opposition to the divine right of states. With the return of southern influence through the democratic party the principle of divine right in the states was protected in an emphasis on the divine right of the United States expressed in such intangibles as a way of life. The pattern of federal government in the United States was followed by members of the Commonwealth notably in Canada and Australia.

The reaction of the United States and members of the Commonwealth in their attempts to protect fundamental law has left them more imperialistic than the mother country. As we have traced the reassertion of common law in Great Britain and the decline of imperialism we must turn to its decline in the other Anglo-Saxon regions and the right of imperialism. In the English colonies in North America which became the United States, rights were protected in the constitution. Control over land within the boundary of each state remained in control of the state but beyond the boundary of the coastal states in the interior of the continent it was in the hands of federal authorities until a new state was set up and accepted in the union. Expansion across North America proceeded to the Pacific Coast and new systems of

---

[1] Brooks Adams, The Emancipation of Massachusetts, the dream and the reality (Boston, 1919).

control were developed beyond the borders in Alaska, Hawaii, the Philippines and other areas. It has been said that the British Empire was acquired in a fit of absent-mindedness, but the American Empire has grown up during periods of imperialistic fanaticism marked by such slogans as *Manifest Destiny* and *54-40 or fight*, and during periods when imperialism was thrust upon her as in the Louisiana Purchase. In Canada we have seen the devices at work in various forms ranging from the fisheries disputes to protests against construction of the Canadian Pacific Railway and the duress exercised by President Theodore Roosevelt on the arbitrators in the Alaska boundary dispute. Significantly other countries are beginning to see the character of American imperialism. American publications protest against appointments of certain cabinet members in Great Britain. An American public body passed a resolution demanding the settlement of the Irish question. Shades of George III! It has been largely in response to the pressure from American imperialism that Canada has developed her own type of imperialism. Nova Scotia entered Confederation on a condition that the resources of the larger federal unit should be used to compel the United States to recognize her rights in the fisheries. Canada has no hesitation in using her influence to prevent a treaty between Newfoundland and the United States which seemed to threaten her bargaining position in the fisheries. The Act of Union was designed to enable Ontario and Quebec to develop transportation facilities which would meet American competition. Expansion of Confederation westward was designed to check encroachments from the United States. The policy of the Dominion in the development of the Prairie provinces was evident in the support of the Canadian Pacific Railway and in land policies designed to check American aggression. In resisting American imperialism we developed our own type of imperialism: its character became evident in the growing insistence on nationalism shown in the defeat of the reciprocity treaty, in the peace treaty, in the Statute of Westminster and finally in the acquisition of Newfoundland. It would not be difficult to collect a series of slogans comparable to those of the United States illustrating our imperialistic ambitions. Fittingly enough they might begin with the comment made at the beginning of the century, "The twentieth century is Canada's." In the United States the shift from an obsession with domestic concerns to foreign policy becomes apparent towards the end of the last century. The isolationism of Washington was replaced by the imperialism of McKinley; but it was an imperialism with a bad conscience and of unbelievable crudity to refer again to the tactics of

Theodore Roosevelt not only in the Alaska boundary dispute but also in the Panama Canal negotiations. It was perhaps best expressed in the phrase attributed to him, "What is the constitution between friends?" Conscience reasserted itself in the reduction of tariffs on newsprint after the reciprocity treaty was defeated by Canada in 1911 and in the repeal of measures designed to improve the position of other powers especially Great Britain in the use of the Panama Canal. Rejection of the reciprocity treaty by Canada was a protest against crude imperialism as was to some extent the defeat of the Republican party in the United States. The election of Wilson, the reluctance to become embroiled in the first world war, the lofty sentiments expressed by Wilson on the entry of the United States in the first world war and the refusal to accept the League of Nations were evidence of an uneasiness about imperialistic tendencies. Such uneasiness proved in itself however to be a spur to further imperialistic concern. Loans to European countries were interpreted as debts and consequently as subject to the payment of interest and ultimate repayment. In the words of President Coolidge, "They hired the money, didn't they?" Insistence on recognition of debts strengthened the plea of debtors for loans from the United States with which interest on debts to the United States could be paid. The burden of reparations on Germany was met by various devices in Germany and without, ranging from inflation to the expedients of the Young and the Dawes plans. The great merry-go-round which began with President Harding's interest in normalcy ended with President Hoover's earnest statement that the world was in a new financial era and that technological advance was such that it could support indefinite improvement in standards of living. Unhappily not even presidential assurances were sufficient to prevent the financial crash of 1929 and the consequent depression. The whole elaborate house of cards collapsed. Great Britain went off the gold standard, Hitler came into power and Roosevelt II became President. Uneasy imperialism or uneasy isolation had not paid off. Consequently the depression was marked by a return to isolationist and domestic policies. Roosevelt II without acknowledgement to Thoreau proclaimed that the only fear we have to fear is fear. The United States was concerned with legislation designed to protect her from foreign entanglements. Isolationist policies had been evident in high tariffs notably the Hawley Smoot tariff and had compelled counter measures in other countries notably the Ottawa agreements of the British Commonwealth. During the period of retreat Hitler began a programme of rapid expansion in Germany paralleled to some extent by a similar programme of Mussolini in Italy and by attacks on Manchuria from Japan. Great Britain became involved in a long series of manoeuvres ranging from the abdication of Edward VIII and the visit of the King and Queen to the meetings in Munich designed to delay the inevitable struggle, and to prepare with all possible energy during the delay, notably by impressing on North America a reluctance to engage in war and a determination to become involved only on extreme provocation. The results scarcely need to be detailed since we are much too familiar with the history of the war and the phases leading to our present discontents. Lessons had been learned in the first world war of which full advantage was taken in the second world war. Systems of controls had been worked out during the long period of preparation after 1934 and were immediately applied on the outbreak of war. Devices elaborated in Canada were used by American propagandists as illustrations of possible improvement in American controls with the result that Canadians reading the literature of American propagandists obtained a very superior picture of their superior virtues. In the United States the dangers of large loans to allies were avoided by the ingenious system of lend lease. As a result of the applications of the lessons of the first world war the peace has been characterized by new developments. Fear of Germany in the east and the west following two world wars has prevented the signing of peace treaties and left that country divided between various interests. Fear of a depression during a possible reconversion period from war to peace which followed the first world war until the system of American loans for repaying American debts was devised has favoured an emphasis on military expedients ranging from the Marshall Plan to the Atlantic Pact by which full employment can be assured. Militarism becomes a necessity to the continued export of goods and to continued employment. The emphasis on communism has been an important element in persuading Americans that they must buy their own business. It would be unwise for me to comment on American foreign policy but perhaps you will allow me to quote from American writers. Archibald MacLeish in an article on "The Conquest of America" in the August number of the *Atlantic Monthly* [1] writes, "Never in the history of the world was one people as completely dominated intellectually and morally by another as the people of the United States by the people of Russia in the four years from 1946 through 1949. American foreign policy was a mirror image of Russian foreign policy. Whatever the

---

[1] 1949.

Russians did, we did in reverse." H. Ickes in the *New Republic* [1] wrote "we have been subjugated by Russia because of our fear of Russia." "I thank God that Roosevelt is not here now to see a greater and a stronger America not on its knees but on its hands and knees grovelling before dangers of its own imagining." The outside can perhaps see more clearly than these writers the truth of their remarks in the work of the Committee on Un-American Activities, in the reign of terror introduced as a result of a revival of a system of informers in ex-communists' rackets, in trials and penalties and in rumours of suicides such as one heard in the stories from Germany and Italy. Bertrand Russell has described totalitarian countries as condemning people to lives of perpetual enthusiasm. In turn we seem to be condemned to lives of perpetual hate.

Repercussions of these developments have been strikingly evident in academic life in Canada. If a member of a staff of a Canadian institution wishes to take advantage of even a temporary appointment in the United States he must choose his relatives and his friends with much greater care than an American citizen. Presumably he must not belong to a party such as the C. C. F. or be involved in any discussions which might make him suspect as a threat to the American way of life. A Canadian citizen may not only be refused admission to the United States but the fact may be drawn forcibly to the attention of the public in American publications. Freedom of speech and of the press has not only been weakened directly as a result of American influence but indirectly as Canadians yield to the acceptance of standards imposed by the United States. The academic world will not overlook an attempt to humiliate its most brilliant scholars by American immigration officials nor will Canadians tolerate affronts to their pride at its most sensitive point. Freedom has been perceptibly narrowed in Canada as a result of American hysteria. In 1950, the middle of the twentieth century, a holy year, surely the lowest ebb in any civilization has been reached when it is possible to threaten the lives of thousands of people with atomic bombs, with scarcely a protest in the interests of common humanity. Fortunately we can still turn to Great Britain and Europe. Scholars turned back at the American border have felt much satisfaction at being given honorary degrees by British universities. But everyone must be disturbed by the appearance of the problem of the American refugee. The imposition of oaths for teachers has involved profound

---

[1] October 17, 1940.

disturbances to American academic life and led to a concern of American scholars in appointments outside the United States. The dangers of using militarism as a device for maintaining full employment shown in American policy as a mirror image of Russian policy are shown more sharply in a mirror image of Russian policy such as we have in Canada. Ideologies are the fig-leaves of militarism. T. S. Eliot has referred to "a true satellite culture as one which for geographical and other reasons, has a permanent relation to a stronger one," [1] and to the reasons against consenting to its complete absorption into the stronger culture. The first "it is the instinct of every living thing to persist in its own being"; the second "that the satellite exercises a considerable influence upon the stronger culture; and so plays a larger part in the world at large than it could in isolation." "The survival of the satellite culture is of very great value to the stronger culture." [2] He proceeds to suggest "that both class and region by dividing the inhabitants of a country into two different kinds of groups leads to a conflict favourable to creativeness and progress"—a point emphasized almost two centuries ago by David Hume. "I do not approve the extermination of the enemy; the policy of extermination or, as is barbarously said, liquidating enemies is one of the most alarming developments of modern war and peace; from the point of view of those who desire the survival of culture. One needs the enemy.... The universality of friction is the best assurance of peace." [3]

I have ventured to digress in these remarks as a means of suggesting that my criticism of the United States and of Canada is intended to be in the interests of both and to protest against a policy of American militarism which compels dependence on the United States. The distortions of the Canadian mirror may be more clearly seen if I describe in more detail the process by which what is called light is reflected. I need only remind you of the influence of American publication on Canadian books, and of the fact that 60 per cent of the circulation of periodicals is dominated by Americans, a reduction from 80 per cent of a couple of decades ago, but a reduction offset to an important extent by the influence of radio broadcasting to be supplemented shortly by television. The rapid advance of technology in the field of communication and the vast American market make it inevitable that the

---

[1] Notes towards the Definition of Culture (London, 1949) p. 54.
[2] Ibid., p. 55.
[3] Ibid., p. 59.

United States should dominate English culture in Canada and that it should exercise a powerful influence on French culture even though the latter is protected by language. One might almost conclude that the Canadian mirror is the American mirror but it is rather a reflection in a smaller mirror in which the American image is sharply focussed. If the American people have been described as "on its hands and knees grovelling before danger of its own craven imagining," the Canadian people might be described as standing on their heads. The most significant indication was the size of the liberal majority in the last election. No satisfactory explanation of this phenomenon based on the assumption that Canadians act rationally has been forthcoming. It has been argued that the Liberals showed themselves to be far more competent in handling election campaigns, that Mr. Drew alienated support by his application of provincial antics to the federal field, that elation over the retirement of the Rt. Hon. William Lyon MacKenzie King spurred Liberals to a new pitch of enthusiasm and so on, but these are not adequate and are scarcely sufficient to explain why the electorate felt that a strong opposition was not important. It may be suggested that militarism played its role in that emphasis was given to it by all parties and that such emphasis could have no other effect than strengthening the party in power. Nothing is more ominous than the facility with which the tendency toward totalitarianism has enabled governments to create and exploit crises particularly in periods preceding elections. Mr. Churchill's genius as a politician in the British elections was evident in his recognition of this fact shown in the popularity of his proposal for a discussion of the problem of cold war at top levels. The threat of communism was stressed by the Conservatives as a means of smearing the C. C. F. In turn the C. C. F. was compelled to stress its reactionary characteristics in order to evade criticism. The weakness of smaller parties evident in their tactics became a source of strength to the Liberals. As a result the political shape of Canada began to assume characteristics similar to those of Russia. The rise of a politburo in Canada comparable and paralleling that of Russia effectively diverts attention to its character by pointing to the dangers of the politbureau in Russia. The distortion of Canadian political life has been evident in the attempts of the ambitious to acquire prestige by exploiting Russian stupidity. The stupidity of Russians inciting the attacks of ambitious Canadian leaders has been paralleled by the stupidity of Canadians in recognizing the incitement. In the field of labor the distortion has been evident in the hardening of labour organizations following much publicized purges of communists, by a more rigid discipline, a greater capacity to exact demands and a greater determination to carry out their plans. In Canada a powerful bureaucracy, in part a product of bilingualism, built up in the depression and during the war, continued to exercise a powerful influence in a period of peace to an important extent by insisting that war had never ceased. Centralization which developed rapidly during the depression and was accompanied by a strong civil service and a decline of cruder forms of patronage was followed by the growth of provincial autonomy parties. The stupefying effects of the bureaucracy have been partly a result of the problem of a dual language in government and administration which blunts political edges. Mr. King as Prime Minister emphasized the importance of a French partner but his successor Mr. St. Laurent has no single individual who can take the place of Mr. King as an English partner. He has rather a group of younger English members of the Cabinet anxious ultimately to assume his mantle. The technique of Mr. King of eliminating rivals at the appropriate time has been to some extent denied his successor. Of more serious consequence has been the destruction of our sense of humour which has accompanied a lack of sense of proportion and a lack of criticism. No one can be a social scientist in Canada without a sense of humour. I offer this remark as a footnote to an understanding of Stephen Leacock. The appointment of the President of the Canadian National Railway because he had been deputy governor of the Bank of Canada and had built up prestige in the Wartime Prices and Trade Board by violating the traditions of anonymity in the civil service has created no ripple of amusement throughout Canada. But perhaps I have been forced to concentrate too much on Ottawa papers. Within the space of a week or so he appeared as an authority on trade, banking, combines and railways. In the words of Anita Loos: "A joke is a joke but no one wants to die laughing." The hazards of our profession are becoming serious.

The results of an overwhelming majority in the federal government and of control by the Liberal Party of the Senate and the bench have been evident in various directions. It has left individual provinces as the only opposition, enabled the premier of a province to become the Conservative Leader of the Opposition, and accentuated the problem of federal government. Parties other than the Liberal party tend to dominate the provinces. Consequently dominion-provincial relations occupy a more important role in Canadian politics. Development of opposition from labour and the C. C. F. in some provinces has been followed by coalitions of liberals and conservatives. General disequilibrium and instability have necessitated enhancement of the power of the dominion evident in abolition of appeals to the privy council and in attempts to develop formulae for amendments to the constitution. The tendency towards centralization has accentuated an interest in defense and the creation of an impasse strengthening the influence of the United States. The sense of omnipotence derived from an emphasis on the theory of the divine right of legislatures developed in the federal government compels a sense of omnipotence in provincial governments and it is no accident that the Province of Ontario outraged a sense of justice by retroactive legislation and that the federal government created a sense of futility by disregarding its own regulations in the Department of Justice in dealing with the Combines Report on flour milling. The divine right of legislatures has contributed to the breakdown of the federal structure. Destruction of political relations between the parties of the federal government and those of many of the provinces has widened the gap between the provinces and the Dominion. A decline in the practice of the federal government of recruiting politicians from the provinces and resort to that of building up the federal cabinet from federal politicians have sharpened the differences between the provinces and the dominion. The problem has become more acute as a result of increased emphasis on central monetary policy. The basis of federalism in which the provinces maintained or acquired control over natural resources has been largely destroyed as a result of an increasing emphasis on monetary policy and particularly on large scale resort to income taxes. Provinces and municipalities have been compelled to rely to an increasing extent on other taxes and control of the federal government has been strengthened by division of powers and decline of the principle of taxation without representation. Decline of the principle of taxation without representation has implied resort to agreements and large scale arrangements for transfers between regions. Conflicts arising from the dependence of regions on European markets and of other regions on American markets and the political power of the densely populated regions dependent on the United States compels resort to political patronage on a large scale to areas less effectively represented. Federal patronage has been essential to the prosperity of agriculture in Western Canada. The extreme complexity of government and the inability of the average citizen to understand its problems increases the responsibility of the bureaucracy. The latter are compelled to insist on democracy as a means of hiding the necessity of working contrary to democratic principles. In turn scepticism, such as indicated in this paper, of discussions of democracy are inevitable. The franchise has been extended, redistribution carried out with due regard to the advantages of the party in power, and large numbers have been appealed to by the parties concerned—all designed to strengthen democracy and calculated to work out to the advantage of the bureaucracy. The great art of political success dependent on keeping Scottish Presbyterians and French Canadians in the same party is no longer necessary. It is impossible in this paper to discuss exhaustively the effects of the enormous majority of the Liberal Party in Canadian life. Politics can no longer be discussed in terms of principles and with reference to abstractions. The power of the bureaucracy precludes an appeal to principles and compels concentration on details. Effective criticism becomes impossible with the deliberate attempt to focus attention on external affairs and emphasis on the necessity of presenting a unified front to the point that essential control over military matters, regarded as the essence of sovereignty, is geared to the United States. There is still a fable to the effect that supping with certain mythological figures should only be done with a long spoon. We can appreciate the words of James Fitzjames Stephen "'Le self government', which not infrequently means the right to misgovern your immediate neighbours without being accountable for it to any one wiser than yourself." [1]

It may be argued that all these problems will be solved by the abolition of appeals from the Supreme Court to the Privy Council. I have referred elsewhere [2] to the important position of the legal profession in Canadian politics and it becomes necessary to consider the problem of law at greater length. Dicey has remarked that "federalism substitutes litigation for legislation" and if we are to understand the prospects of success of the federal system we must pay some attention to the nature of the body before which litigation is carried out.

The extent to which the new powers of an enlarged Supreme Court may be able to solve the problems of a federal state will engage the attention of citizens concerned with continuation of the traditions of common law. Federal constitutions provide hiding places for vested interests. The rights of property entrenched in written constitutions restrict possible developments of socialism such as have been evident in Great Britain. Sharp differences emerge between business and government. In federal constitutions emphasizing the traditions of Roman law in common law countries Supreme Courts occupy a crucial posi-

---

[1] Liberty, Equality, Fraternity (London, 1874) p. 268.
[2] Great Britain, the United States and Canada. (Nottingham, 1948)]

tion. Common law traditions assume that the state is part of the law and the subject has greater difficulty in separating himself from the state. Change is consequently more gradual and less subject to revolution. Constitutions are largely protected from drastic revision. But Roman law tradition favoured by written constitutions in the United States and in members of the Commonwealth lean toward imperialism, and threaten the beneficient effects of common law in Western civilization. Without a recognition of the flexibility of common law the remark of Dean Pound that "legal precepts are almost certain to lag behind public opinion whenever the latter is active and growing" will become extremely pertinent. These fundamental problems face the Canadian Courts and the Canadian people. As a result of a firm belief in the impossibility of the spread of communism in common law countries and in the danger of American imperialism in exploiting us through its propaganda about communism I have felt compelled to seize this opportunity to describe our difficulties. The sense of terror which has seized on Canadian life has made it more imperative that I should regard the 150th anniversary of the University of New Brunswick as an occasion on which our faith in the traditions of common law, which were reflected after the American revolution in the founding of this university, could be reaffirmed.

Also from Benediction Books …
Civilization And Its Discontents
Sigmund W. Freud
Benediction Classics, 2011
148 pages
ISBN: 978-1-84902-274-3

Available from www.amazon.com, www.amazon.co.uk

This is Sigmund Freud's seminal text where he explores what he sees as the fundamental tensions between civilization and the individual. He contends that the greatest friction is between the individual's quest for freedom and civilization's contrary demand for conformity. Thus humankind's primitive instincts to kill and sexual gratification are harmful to human society, so we create laws and severe punishments to prevent murder, rape, and adultery. Freud argues that this results in perpetual feelings of discontent in its citizens.

The Book of Khalid
Ameen Fares Rihani
Benediction Classics, 2011
246 pages
ISBN 978-1-84902-424-2

Available from www.amazon.com, www.amazon.co.uk

The Book of Khalid is a novel that was written in 1911 by the Arab author Ameen Fares Rihan. He wrote in English and described the perilous journey of immigrants to America. This was the first book to paint a picture of immigration, and also the first to break the barrier between East and West. In telling the story it points out the conflicts between Eastern and Western values and culture and attempts to reconcile them. This edition comes with the original illustrations by his friend and colleague Khalil Gibran, and some think that this book influenced "The Prophet".

Eureka: A Prose poem. An Essay on the Material and Spiritual Universe.
Edgar Allan Poe
Benediction Classics, 2011
100 pages
ISBN: 978-1-78139-037-5

Available from www.amazon.com, www.amazon.co.uk

Eureka, A Prose Poem is over 40,000 words in length and it is not a poem! It was Poe's last major work and was based on a lecture Poe presented in 1848, titled "On The Cosmography of the Universe". He had hoped for an audience of hundreds and that the proceeds of the lecture would pay for his new journal "The Stylus". However, only 60 attended and they went away confused. In Eureka, Poe explains the universe and the relationship of people with God based on the proposition "Because Nothing was, therefore All Things are".

How The Gods Were Made (A study in Historical Materialism). In original format.
John Keracher
Oxford City Press, 2011
66 pages
ISBN: 978-1-78139-030-6

Available from www.amazon.com, www.amazon.co.uk

A classic reprint of a 1929 text, explaining the materialist arguments against gods. This gives a clear explanation of the Socialist opposition to religion.

Also from Benediction Books …
Wandering Between Two Worlds: Essays on Faith and Art
Anita Mathias
Benediction Books, 2007
152 pages
ISBN: 0955373700

Available from www.amazon.com, www.amazon.co.uk

In these wide-ranging lyrical essays, Anita Mathias writes, in lush, lovely prose, of her naughty Catholic childhood in Jamshedpur, India; her large, eccentric family in Mangalore, a sea-coast town converted by the Portuguese in the sixteenth century; her rebellion and atheism as a teenager in her Himalayan boarding school, run by German missionary nuns, St. Mary's Convent, Nainital; and her abrupt religious conversion after which she entered Mother Teresa's convent in Calcutta as a novice. Later rich, elegant essays explore the dualities of her life as a writer, mother, and Christian in the United States-- Domesticity and Art, Writing and Prayer, and the experience of being "an alien and stranger" as an immigrant in America, sensing the need for roots.

About the Author

Anita Mathias is the author of *Wandering Between Two Worlds: Essays on Faith and Art*. She has a B.A. and M.A. in English from Somerville College, Oxford University, and an M.A. in Creative Writing from the Ohio State University, USA. Anita won a National Endowment of the Arts fellowship in Creative Nonfiction in 1997. She lives in Oxford, England with her husband, Roy, and her daughters, Zoe and Irene.

Anita's website:
   http://www.anitamathias.com, and
Anita's blog Dreaming Beneath the Spires:
   http://dreamingbeneaththespires.blogspot.com

The Church That Had Too Much
Anita Mathias
Benediction Books, 2010
52 pages
ISBN: 9781849026567

Available from www.amazon.com, www.amazon.co.uk

The Church That Had Too Much was very well-intentioned. She wanted to love God, she wanted to love people, but she was both hampered by her muchness and the abundance of her possessions, and beset by ambition, power struggles and snobbery. Read about the surprising way The Church That Had Too Much began to resolve her problems in this deceptively simple and enchanting fable.

About the Author

Anita Mathias is the author of *Wandering Between Two Worlds: Essays on Faith and Art*. She has a B.A. and M.A. in English from Somerville College, Oxford University, and an M.A. in Creative Writing from the Ohio State University, USA. Anita won a National Endowment of the Arts fellowship in Creative Nonfiction in 1997. She lives in Oxford, England with her husband, Roy, and her daughters, Zoe and Irene.

Anita's website:
    http://www.anitamathias.com, and
Anita's blog Dreaming Beneath the Spires:
    http://dreamingbeneaththespires.blogspot.com

The Origin Of Species By Means Of Natural Selection; Or The Preservation Of Favoured Races In The Struggle For Life (Sixth Edition, with all additions and corrections)
Charles Darwin
Benediction Classics, 2011
482 pages
ISBN: 978-1-84902-472-3

Available from www.amazon.com, www.amazon.co.uk

The complete sixth edition of Darwin's classic, with all additions and corrections.

Collected Essays, Volume 2, Darwiniana
Thomas Henry Huxley
Benediction Classics, 2011
296 pages
ISBN: 978-1-84902-403-7

Available from www.amazon.com, www.amazon.co.uk

This book contains the eleven essays by Huxley on evolution. Huxley was known as 'Darwin's Bulldog', so called because he was a great advocate of Darwin's evolution theory. He was also made famous as a result of his debate with Samuel Wilberforce in Oxford on 1860, a debate caused wider acceptance of the theory of evolution. This book comes complete with a preface by Huxley and some black and white drawings. The essays included are as follows: The Darwinian Hypothesis [1859] The Origin of Species [1860] Criticisms on "The Origin of Species" [1864] The Genealogy of Animals [1869] Mr. Darwin's Critics [1871] Evolution in Biology [1878] The Coming of Age of "The Origin of Species" [1880] Charles Darwin [1882] The Darwin Memorial [1885] Obituary [1888] Six Lectures to Working Men "On Our Knowledge of the Causes of the Phenomena of Organic Nature" [1863]

**Gomez Arias;
Or The Moors Of The Alpujarras.
A Spanish Historical Romance
Joaquín Telesforo De Trueba Y Cosío**
Benediction Classics, 2011
354 pages
ISBN: 978-1-84902-462-4

Available from www.amazon.com,
www.amazon.co.uk

This paperback includes all three Volumes of the novel 'Gomez Arias The Moors of the Alpujarras, A Spanish Historical Romance'. It was written in 1826 by the Spaniard Joaquín Telesforo de Trueba y Cosío who wrote in English and belonged to the Romantic movement. It is set in the province of Granada. A censored version was produced in 1831

**Crestlands - A Centennial Story of Cane Ridge - with illustrations
Mary Addams Bayne**
Benediction Classics, 2011
228 pages
ISBN: 978-1-84902-325-2

Available from www.amazon.com,
www.amazon.co.uk

Crestlands, A Centennial Story of Cane Ridge is a romance, set in Kentucky during the early part of the nineteenth century. The story gives a fascinating portrayal of the life and conditions for the people at this time. The book comes complete with original illustrations.

Life of an American Workman
Walter P. Chrysler
Oxford City Press, 2011
156 pages
ISBN: 978-1-84902-395-5

Available from www.amazon.com, www.amazon.co.uk

Walter Chrysler established the Chrysler Corporation from the embers of the Maxwell-Chambers Auto Works; then founded the much larger Dodge Brothers Motor Co. In this book he tells his story simply and with many entertaining anecdotes. He talks about his childhood, his work on the railroads, his turn-around of American Locomotive, Buick and then Maxwell-Chambers. Chrysler writes in an easy style, in the book he explains his philosophy that R&D are vital to a company's success. The Chrysler Building was in the planning stage and that is discussed in the book along with his belief that his children must start at the bottom and worth their way up. This easy-to-read autobiography is a delight for anyone but a must-read for automotive enthusiasts.

Woman: Her Charm and Power
Robert P. Downes
Benediction Classics, 2011
366 pages
ISBN: 978-1-78139-009-2

Available from www.amazon.com, www.amazon.co.uk

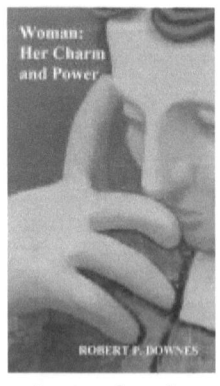

This is a book written in 1900, describing the status and treatment of women. It is a culturally important book, as well as being both interesting and a pleasure to read. This book shows in stark contrast how far women have come since it was first published.

Lincoln's Last Hours, Memorial Address On The Life And Character Of Abraham Lincoln, The Perfect Tribute, Abraham Lincoln, A Memorial Discourse
Charles A. Leale
Benediction Classics, 2011
106 pages
ISBN: 978-1-84902-405-1

Available from www.amazon.com, www.amazon.co.uk

LINCOLN'S LAST HOURS
by Charles A. Leale

MEMORIAL ADDRESS ON THE LIFE AND CHARACTER OF ABRAHAM LINCOLN
by George Bancroft

THE PERFECT TRIBUTE
by Mary Raymond Shipman Andrews

ABRAHAM LINCOLN, A MEMORIAL DISCOURSE
by Rev. T. M. Eddy, D. D.

This is a collection of four historic tributes to President Abraham Lincoln:
1. LINCOLN'S LAST HOURS by Charles A. Leale Address delivered before the commandery of the state of New York Military order of the loyal legion of the United States at the regular meeting, February, 1909, city of New York in observance of the one hundredth anniversary of the birth of President Abraham Lincoln.
2. MEMORIAL ADDRESS ON THE LIFE AND CHARACTER OF ABRAHAM LINCOLN by George Bancroft. Delivered, at the request of both Houses of the Congress of America, before them, in the House of Representatives at Washington, on the 12th of February, 1866.
3. THE PERFECT TRIBUTE by Mary Raymond Shipman Andrews Written in 1908, a story paying tribute to the character of Abraham Lincoln.
4. ABRAHAM LINCOLN, A MEMORIAL DISCOURSE, By Rev. T. M. Eddy, D. D., Delivered at a Union Meeting, held in the Presbyterian Church, Waukegan Illinois, Wednesday, April 19, 1865, the day upon which the funeral services of the president were conducted in Washington, and observed throughout the loyal states as one of mourning.

www.ingramcontent.com/pod-product-compliance
Ingram Content Group UK Ltd.
Pitfield, Milton Keynes, MK11 3LW, UK
UKHW041448180426
11946UKWH00001B/5